THE NEW LOCALISM

OTHER RECENT VOLUMES IN THE SAGE FOCUS EDITIONS

8. **Controversy (Third Edition)**
Dorothy Nelkin
41. **Black Families (Second Edition)**
Harriette Pipes McAdoo
64. **Family Relationships in Later Life (Second Edition)**
Timothy H. Brubaker
89. **Popular Music and Communication (Second Edition)**
James Lull
133. **Social Research on Children and Adolescents**
Barbara Stanley and Joan E. Sieber
134. **The Politics of Life in Schools**
Joseph Blase
135. **Applied Impression Management**
Robert A. Giacalone and Paul Rosenfeld
136. **The Sense of Justice**
Roger D. Masters and Margaret Gruter
137. **Families and Retirement**
Maximiliane Szinovacz, David J. Ekerdt, and Barbara H. Vinick
138. **Gender, Families, and Elder Care**
Jeffrey W. Dwyer and Raymond T. Coward
139. **Investigating Subjectivity**
Carolyn Ellis and Michael G. Flaherty
140. **Preventing Adolescent Pregnancy**
Brent C. Miller, Josefina J. Card, Roberta L. Paikoff, and James L. Peterson.
141. **Hidden Conflict in Organizations**
Deborah M. Kolb and Jean M. Bartunek
142. **Hispanics in the Workplace**
Stephen B. Knouse, Paul Rosenfeld, and Amy L. Culbertson
143. **Psychotherapy Process Research**
Shaké G. Toukmanian and David L. Rennie
144. **Educating Homeless Children and Adolescents**
James H. Stronge
145. **Family Care of the Elderly**
Jordan I. Kosberg
146. **Growth Management**
Jay M. Stein
147. **Substance Abuse and Gang Violence**
Richard E. Cervantes
148. **Third World Cities**
John D. Kasarda and Allan M. Parnell
149. **Independent Consulting for Evaluators**
Alan Vaux, Margaret S. Stockdale, and Michael J. Schwerin
150. **Advancing Family Preservation Practice**
E. Susan Morton and R. Kevin Grigsby
151. **A Future for Religion?**
William H. Swatos, Jr.
152. **Researching Sensitive Topics**
Claire M. Renzetti and Raymond M. Lee
153. **Women as National Leaders**
Michael A. Genovese
154. **Testing Structural Equation Models**
Kenneth A. Bollen and J. Scott Long
155. **Nonresidential Parenting**
Charlene E. Depner and James H. Bray
156. **Successful Focus Groups**
David L. Morgan
157. **Race and Ethnicity in Research Methods**
John H. Stanfield II and Rutledge M. Dennis
158. **Improving Organizational Surveys**
Paul Rosenfeld, Jack E. Edwards, and Marie D. Thomas
159. **A History of Race Relations Research**
John H. Stanfield II
160. **The Elderly Caregiver**
Karen A. Roberto
161. **Activity and Aging**
John R. Kelly
162. **Aging in Rural America**
C. Neil Bull
163. **Corporate Political Agency**
Barry M. Mitnick
164. **The New Localism**
Edward G. Goetz and Susan E. Clarke
165. **Community-Based Services to the Elderly**
John A. Krout
166. **Religion in Aging and Health**
Jeffrey S. Levin
167. **Clinical Case Management**
Robert W. Surber
168. **Qualitative Methods in Aging Research**
Jaber F. Gubrium and Andrea Sankar

Alan C. Brown.

THE NEW LOCALISM

Comparative Urban Politics in a Global Era

Edward G. Goetz
Susan E. Clarke
editors

SAGE PUBLICATIONS
International Educational and Professional Publisher
Newbury Park London New Delhi

For information address:

SAGE Publications, Inc.
2455 Teller Road
Newbury Park, California 91320

SAGE Publications Ltd.
6 Bonhill Street
London EC2A 4PU
United Kingdom

SAGE Publications India Pvt. Ltd.
M-32 Market
Greater Kailash I
New Delhi 110 048 India

Printed in the United States of America

Library of Congress Cataloging-in-Publication Data

Main entry under title:

The new localism: comparative urban politics in a global era / edited by Edward G. Goetz, Susan E. Clarke.
p. cm. — (Sage focus editions; 164)
Includes bibliographical references and index.
ISBN 0-8039-4921-9 (cloth).—ISBN 0-8039-4922-7 (pbk.)
1. Local government. 2. Economic development. I. Goetz, Edward G. (Edward Glenn), 1957- . II. Clarke, Susan E., 1945- .
JS91.N5 1993
320.8—dc20 93-11773
CIP

93 94 95 96 97 10 9 8 7 6 5 4 3 2 1

Sage Production Editor: Yvonne Könneker

Contents

1

The New Localism

Local Politics in a Global Era

SUSAN E. CLARKE

It seems incongruous to foresee a new localism in an era of global economic change. It may appear even more fanciful to tie localism trends to democratic reform processes, especially in settings as disparate as Kenya, Nigeria, Poland, Hungary, the Caribbean, the United Kingdom, and the United States. For one thing, emergent theories about the local effects of global restructuring processes are partial and not particularly robust. And the notion of democratic reform seems salient for countries undergoing highly visible democratic transitions, but less so for settings such as Africa, where transitions are less visible and even more fragile (Holmquist & Ford, 1992). Even more puzzling to some is the application of democratic reform concepts to established political systems in advanced industrial societies.

Nevertheless, there is some merit in such an application. Global economic change processes generally transcend scale, yet in some instances are very sensitive to local contextual factors, including state actions. Recent evidence of intensified economic and political demands on localities and increased local development initiatives hints at a new local terrain with unexpected commonalities. Local officials the world over operate under heightened conditions of economic and political uncertainty. They now have social and economic roles and responsibilities that are often new and unanticipated. In each instance, global restructuring pressures compel local officials to reconstruct relations between the public and private sectors at the local level as well as to reconsider the most basic governance issues.

This volume draws not only on European and American experiences but also on studies of Central Europe, the Caribbean, and Africa. It brings together analyses of local political restructuring by urban scholars active in both research and policy debates on these questions. Some contributors to this volume presented earlier versions of these studies at panels organized by Terry Nichols Clark for the International Sociological Association's biennial meeting in Madrid in 1990.[1] Complementary papers came to our attention at other professional conferences. Our objectives in presenting them here are threefold: first, to address an intriguing phenomenon at the heart of recent analyses of comparative urban politics—the emergence of local policy activism, particularly economic development activism, in the face of countervailing trends toward a global economy; second, to integrate urban political economy theories of scale-sensitive economic processes with theories of democratic transition that tend to overlook scale effects; and third, to describe local political restructuring processes under way in response to these economic trends and pressures for democratic reform.

The New Localism

One of the more influential interpretations of global economic restructuring depicts local actors with little room for maneuvering (Piore & Sabel, 1984; Sassen, 1990). In this view, the hypermobility of capital pits community against community in competition for private investment and limits the abilities of nation-states and localities to carry out autonomous economic and social policies. The policy implications of this perspective have been momentous. Many national governments acted on this interpretation of restructuring in ways that trivialized the roles of cities in the national economy (Barnekov, Boyle, & Rich, 1989). They advocated the need for enhanced national competitiveness and viewed any spatially targeted policies as hindering efficient locational investment decisions. Thus specific urban policies and many social policies seen as "unproductive" were abandoned. This has been increasingly so in the United States and Europe, even in countries previously seen as in the forefront of urban and social policy development. These tensions also characterize the policy setting in Central

Europe, the Caribbean, and Africa, where political systems are undergoing tremendous challenge.

There is a growing sense that such totalizing perspectives hamper theoretical development and impede political analysis (Clark, McKay, Missen, & Webber, 1992; Cox, 1992). Thus a critical reassessment of our understanding of the local impacts of globalization is under way. Although there is general agreement on globalization trends, there is less consensus on concomitant economic and political dynamics. This includes, for example, continued debate on whether global economic restructuring involves a transition to a new period of capital accumulation distinct from the "Fordist" era. More particularly, arguments that new flexible production processes are associated with the global organization and hypermobility of capital and a locational logic of both "new industrial spaces" and victimized localities are criticized as overgeneralized (Cox, 1992; Logan & Swanstrom, 1990; Preteceille, 1990) and underspecified (Cox, 1992; Henry, 1992). A number of relationships are called into question: whether there is a necessary relation between hypermobility and globalization (Cox, 1992), between uncertainty and organizational responses to reduce transaction costs through agglomeration (Henry, 1992), between hypermobility and local political change (Cox, 1992), and indeed among flexible specialization, localism, and decentralization policies (Preteceille, 1990). Because a more scale-sensitive and state-sensitive understanding of global restructuring processes remains elusive, the potential economic and political roles for localities in a global era are unclear.

Scale-Sensitive Perspectives on Restructuring Processes

The reassessment of global restructuring perspectives encompasses not only analyses of the varying effects of global economic change at different geographic scales, but also, of special interest here, the differing consequences of societal change for production systems at different scales (Gaspar, 1992, p. 828; Logan & Swanstrom, 1990). The chapters in this volume sketch some of the latter consequences; they underscore our contention that theories of democratic and market transition must be more sensitive to scale effects and that economic restructuring theories must accommodate state-sensitive concepts.

Recognizing that the consequences of societal changes, such as political restructuring processes, for production systems vary at different scales opens up a number of issues regarding the economic and political rationales for a new localism. Overall, the economic rationale for localism stems from the general sense that internationalization and restructuring processes have contributed to a duality of the global and the local (Gaspar, 1992, p. 830). Seen from this perspective, variations in production environments across localities become especially important for restructuring processes and economic performance. Thus the scale effects of economic restructuring processes may suggest enhanced roles for localities nested in "new industrial spaces" created by these processes. Analysts as well as local officials have struggled to identify the locality-specific dimensions of these production environments that allow some communities to gain a competitive edge. One result is a rather disconnected view of localism based on the rise and spread of new industrial spaces that are linked but not vertically integrated. The profusion of attempts to mimic aspects of prosperous areas such as the Silicon Valley and the M4 corridor reflects this perhaps excessive and exclusive emphasis on local features.

More theoretically grounded versions emphasize the links between the logic of new production systems and their spatial dynamics, particularly the local consequences of the need for greater integration and coordination among decentralized sites. To some analysts, these needs in themselves argue for a "return to place" for certain types of industries. Two versions of this argument have important ramifications for local roles: the emphasis on the value of local Marshallian "industrial atmospheres" for certain industries (Amin & Thrift, 1992) and the emphasis on these firms' need to reduce transaction costs (Scott, 1992). Characterizing Marshallian environments in terms of specific local sociocultural features stresses basic structures for growth that may be beyond the scope of local policy initiatives and possible only in a limited number of places (Amin & Thrift, 1992). A focus on transaction costs highlights the need for an institutional infrastructure that facilitates the minimization of transaction costs but hints that these institutional innovations are within the ken of local policy makers. In both instances, locale becomes critical: To the extent that new production processes are, in fact, characterized by organizations seeking specific industrial atmospheres and attempting to reduce external transaction costs (see Henry, 1992), localities—not regions or nations—are in a position to facilitate the minimization of these costs (Mayer, 1989).

State-Sensitive Perspectives on Restructuring Processes

Although these economic and spatial perspectives hint at the political rationale for a new localism, the political relations stemming from these economic changes are undertheorized, remaining too often at the stage of deducing politics and policies from economic and spatial processes (Preteceille, 1990, p. 36). But more decidedly, political analyses suffer from similar if obverse deductive flaws. Typologies of relative community advantage include descriptions of global cities cresting a global urban hierarchy (Logan & Molotch, 1987; Savitch, 1988) as well as more prosaic locales embedded in growth areas. Their relative advantage and potential autonomy are derived from the "changing locational logic of post-fordist flexible production" (Henry, 1992, p. 375), even though communities may not act on these advantages or necessarily have contributed directly to them. In contrast to these efforts to specify the spatial and economic dynamics permitting greater autonomy for a limited number of cities, scores of studies detail local economic development initiatives undertaken independent of these conditions (Clarke & Gaile, 1992; Frieden & Sagalyn, 1989; King & Pierre, 1990; Stohr, 1990; Stone & Sanders, 1987). Although there is little consensus on the defining features of global cities or on the nature and prevalence of urban hierarchies, it remains easier to identify cities embedded in new production systems and enjoying some latitude in policy initiatives than to explain the comparable activism of communities whose hierarchical or network status is less obvious.

Political rationales for localism are not based wholly on privatism or community values or even necessarily locational logics; they also include the instrumental use of localism as a political strategy to circumvent or replace outmoded structures of central bureaucracies. New local institutional arrangements operate in ways that maximize their flexibility, their nonstandardized routines, and their contextual decision rules (Clarke & Gaile, 1989). In contrast to the centrally organized, purposive welfare state, these new arrangements suggest an opportunistic state, one responding pragmatically to opportunities rather than pursuing agendas (Fuchs & Koch, 1991). The rise of the decentrist, opportunistic state implies a flattening of hierarchy (Clark, 1992, p. 8), possibly a political analogue to the webs or networks linking firms in new production processes. But to the extent that the legitimacy of democratic regimes is tied to economic performance rather than governance based on civic values (Fuchs & Koch, 1991), the opportunistic state intimates the subordination of political will to private interests, particularly at the local level.

Overall, despite these insights on local political relations and the greater attention to scale-sensitive perspectives of restructuring processes, urbanists have yet to articulate a persuasive, integrative, theoretical perspective on new localism trends. Rather curiously, this gap is mirrored by theories of democratic reform and regime change that are sensitive to state changes but indifferent to scale.

Low Politics in a Global Era

Moving from theories of localism to local practice draws attention to this theoretical lacuna. Although many comparative urbanists ground their analyses in the context of economic restructuring (Clarke, 1989; Judd & Parkinson, 1989; King & Pierre, 1990; Logan & Swanstrom, 1990; Savitch, 1988), and some comment directly on localism trends (Mayer, 1989; Preteceille, 1990), these analyses have been limited to American and Western European case comparisons. Studies of restructuring localities in transforming societies have been confined, understandably, to individual cases; many are not yet available in English-language versions. This narrow and bifurcated focus has meant that urban scholars have not taken advantage of the frameworks offered by theorists of regime change and democratic reform. And despite the claims of those in transforming societies that they must rebuild "from the bottom up," most theories of democratic reform and regime change fail to incorporate any sense of the role of local political change in these processes. In particular, the indifference to scale means most theories of democratization and marketization assume a nation-state focus and overlook the potentially differential effects of these trends at other geographic scales as well as the effects of locally based activities for national transformations. Borrowing Bialers's terms, we argue that scholarly preoccupation with the "high politics" of restructuring and democratic reform slights the "low politics" involving people's daily lives, local communities, and work (Weigle & Butterfield, 1992, n. 14).

Although we contend that it is not possible to understand regime change and democratic reform fully without greater attention to the role of local politics, the issue here is whether thinking about local political change in terms of democratic reform theories enhances our understanding of variations in local politics in this global era. At first glance, broad theories of democratic reform appear most appropriate for analysis of local politics in societies undergoing dramatic transformations in economic and governance structures. But if we view democratic reforms in more analytic terms, attention would be directed to variations in local features that, for example, institutionalize uncertainty about outcomes, provide for adjudication of continual intergroup conflicts, establish norms for economic and political action, and craft vertical and horizontal institutions that shape capacities of different groups to realize their interests (Przeworski, 1986).

These analytic issues are at the heart of normative debates on local governance across political and economic settings. As one observer points out, Westerners tend to overlook concomitant structural changes in their own societies and thus neglect the need for new theories spanning political change in East and West, North and South (Jorgensen, 1992, p. 650). Keeler (1993) characterizes such theoretical ambitions as the third wave of democratic reform literature in that this wave attempts to compare the politics of democratic reforms across regions and issue areas. Here the goal is to examine local political change in the context of economic restructuring and democratic reform efforts across so-called Third World countries dealing with structural reform programs, transitional democracies in Central Europe, and more established democratic regimes in the United States and Europe.

Political Architecture:
The Restructuring of the Local Political Economy

Reframing the localism issue in terms of the conditions under which changes in the political regulation of the market occur at the local level may offer a more promising path to this third wave of theoretical development than have previous deductive perspectives (Walton, 1990). It requires placing the interrelationship of markets and politics in historical and comparative perspective and directs attention to varying "configurations of market pressures and local capacities for action" (Walton, 1990, p. 254). As described in this volume, the capacity for local action in the face of market reorganization entails the renegotiation of public/private roles and responsibilities, particularly the scope and acceptable forms of state intervention, and the reassessment of central/local arrangements. This process of political restructuring is ongoing: The dramatic transformations in Central Europe, Africa, and the Caribbean accent the conflicting values about these relationships and their institutional expression that also mark debates in countries with more established local government systems. Thus the more overt political architecture under way in transforming societies reintroduces issues of local governance in the United States and Western Europe.

Whereas culturally sensitive understandings of localism will enrich the specific interpretations, the following discussion traces the analytic characteristics of democratization and restructuring processes that play out in the authors' accounts of the new localism. These features include the effects of international organizations on local political change, shifts in central-local relations that alter the capacities of local interests, and changes in local political opportunity structures, including norms that both increase uncertainty about decision outcomes and encourage privatization of local authority and the rise of local public entrepreneurs. In the concluding chapter, Goetz considers these features in the context of cross-sectional data on cities in 12 countries. Underlying these empirical developments are the notions of missed opportunities and preferred alternatives (Przeworski, 1986); metaphors characterizing current realities and preferred alternatives for local political change are noted in the concluding section.

Local Effects of International Opportunity Structures

A new "international opportunity structure" (Tarrow, 1991) is shaping local economic and political changes. In each of the settings discussed here, changes in central-local relations and city-market links are embedded in this larger structure. Transnational capital structures and international organizations such as the International Monetary Fund have strong if indirect effects on local governments across settings in the accounts here; the strategic agendas of international organizations directly affect both political and economic change in local governments in Central Europe, the Caribbean, and Africa.

It is increasingly clear that pressure from international financial organizations accelerated the preferred timing of democratic elections and economic transition in Central Europe. In the Polish and Hungarian cases, the burden of external debt and the inefficiencies of the command economies undermined the legitimacy of the previous regimes to the point where partial reforms were under way by the early 1980s. Initially, debates on these reforms did not presume privatization as an option

(Castle, 1990); as the reforms faltered and oppositional strength compelled political transition, marketization options became prominent features of the debates. Nevertheless, the timing and content of both the political and economic transitions remained contested even after the authoritarian regimes were displaced. The timing and content questions so critical to regime change became conditions for receiving external aid. As a result, local elections were held in Poland before basic governance and financing aspects of central-local relations were resolved (see Surazska, Chapter 5, and Péteri, Chapter 6, this volume; see also Castle, 1990; Fabian, 1992). Newly elected officials did not know whether they owned municipal assets, what their tax bases would be, how much support they would receive from central government, or how much oversight would be exercised over their decisions. As a consequence, the effectiveness of local governments was undermined.

International donor organization pressures were the "trip wire" for democratization reforms in Kenya (Holmquist & Ford, 1992, p. 101). By the mid-1980s, opposition to the one-party rule of President Daniel Arap Moi was increasingly visible and vocal; it was centered in the new urban middle class and voiced its discontent through church groups, lawyers' associations, women's groups, and subtle press reportage of "nonevents." Despite the tremendous personal courage of the people, the "political opening" to democratic reform occurred only when the major donor groups suspended $350 million in foreign exchange relief in November 1991, pending political and economic reforms (Holmquist & Ford, 1992, p. 98). Moi responded in late 1991 by announcing the return of multiparty competition; as Ngau reports in Chapter 9 of this volume, elections were finally set for late 1992, but on such short notice that they were challenged in court and rescheduled. Despite this brief respite, the numerous opposition parties were unable to form an electoral coalition and the president's ticket prevailed, although several cabinet members were defeated. These events address the political conditions set by the donors, but in the absence of any meaningful policy debate; the conditions for such debate atrophied with the clientelistic practices of the one-party era, and the mix of economic and political reform conditions imposed by the donors circumscribed the policy options open to debate (Holmquist & Ford, 1992, p. 98).

The economic effects of international financial organizations on local governments in Poland and Hungary were indirect in that the imposition of national austerity and the need to reduce debt burdens reduced the amount of aid available for local governments. They were more direct in Kenya, Nigeria, and the Caribbean, where structural adjustment programs (SAPs) severely limited the scope of national initiatives and local expenditures. In the Caribbean, SAPs and U.S. trade policies are critical elements in the development schemes formulated by local policy makers, although Klak and Rulli (Chapter 7) point out that the overall external dependency of the region is as important as specific policy initiatives.

Shifts in Central/Local Relations

In almost every instance, the uncertainty over the relative autonomy of local governments is striking. Historical patterns of central-local relations continue to shape emergent structures. In Chapter 6, Péteri details the value conflicts characterizing these debates at the national level; the Nigerian case also illustrates the political and fiscal motivations driving these choices. Surazska, in Chapter 5, indicates that the reintroduction of oversight mechanisms at the regional level may further limit local discretion. Somewhat ironically, the cases of Poland and Kenya suggest that the movement from unitary systems necessitates more territorial tiers of government. Thus democratization trends at the national level have left unclear the autonomy of local governments relative to other tiers.

Yet, in transforming societies this issue rises in the context of relatively weak state organizations at every level. Much of the pressure for devolution in Poland, Hungary, Nigeria, and, to some extent, Kenya stems from the need of the national government to comply with international debt programs and to stave off expenditure responsibilities. Even if the debt burdens were less, there is limited state capacity at the national level to provide necessary resources and administer comprehensive programs. And given the administrative nature of previous local governments, local state capacities are atrophied in Central Europe. The concept of a local state is even more problematic in Africa, where the very juridical integrity of local governments remains at risk (Mabogunje, 1990).

Similar revenue dynamics characterize more established intergovernmental structures in the United States and the United Kingdom: The pressures of budget deficits encourage national governments to cut back on local aid, to issue mandates without operational resources, and to devolve program responsibilities through increasingly hefty matching requirements. In the unitary British system, national governments also attempt to constrain local expenditures and local voice; in the United States, these efforts occur at the state level through tax and expenditure limitations and other constraints on local governance.

New Local Opportunity Structures

To the extent that shifts in the external environment affect capital structures and central-local relations, they help bring about new local political opportunity structures. These political opportunity structures (Eisinger, 1973; Tarrow, 1991) expand and redistribute influence in ways that reflect the economic, social, and political changes under way. These new political opportunity structures will vary, depending on structural features that are relatively endogenous to a particular society: the degree of hierarchy (Clark, 1992), uneven development patterns and regional inequalities, the size of the middle class, independent trade unions, the historical development of civic organizations, and so on. These endogenous factors shape the forms taken by democratization and restructuring trends, but they also affect local adaptability to new demands, roles, and responses to economic changes. Three aspects of contemporary changes in local opportunity structures stand out in these chapters: the extent to which changes in access to influence and uncertainty about outcomes actually occur, the reconstruction of civil society, and the privatization of public authority.

Redistributing local influence. The chapters in this volume suggest that the redistribution of local influence may be problematic. Whereas local political opportunity structures expanded through increased formal political access and disarray in local political alignments (Tarrow, 1991), the responses to these new opportunities are especially influenced by previous domestic capital structures, the role of local parties and political organizations, and the existence of nongovernment associations. It appears, for example, that levels of access to institutional participation have increased in a formal sense, but that participation in local elections remains low. This may reflect practices in former years, where the most political individuals did not participate in elections (Ash, 1990). In addition to the issue of electoral participation, the opportunity to benefit from privatization of local public assets depended heavily on previous opportunities to accumulate domestic capital. In most cases, the party *nomenklatura* enjoyed these opportunities and therefore gained new positions of influence in newly privatized local governments.

Political parties and organizations at the local level in transforming societies are often fragmented and unable to mobilize electoral coalitions. Surazska points out in Chapter 5 that Solidarity's electoral strength varied by the size of place and by the existence of other political parties; in some communities, formerly Communist organizations prevailed. Similarly, recent elections in Kenya highlight the regionality of party organizational strength; although the oppositional parties enjoyed support in urban centers, these new parties were viewed with skepticism and distrust in many rural areas. As Ngau, author of Chapter 9, has pointed out to me, "Our mothers do not know them"; thus national opposition parties foundered on weak local bases and their inability to form coalitions across organizations. At the local level, the inability of new parties to articulate and aggregate local interests hampers local decision making. It is likely that, over time, party organizations will become more effective in mobilization and coalition formation, but they currently offer modest means for response to expanded local political opportunities.

Reconstructing civil society. The evolution of a public sphere or space relatively independent of the state hinges on the ability of associations to establish norms of "publicness." This allows the creation of mediating institutions, permitting individual public participation through associations and providing for state accountability. Given the importance of the historical context in which civil society develops, particularly the character of economic differentiation, it is not surprising that the contributions to this volume indicate that this is an important but highly variable factor in local political change.

To most of the authors, associational activities reflect the extension of the public sphere in ways that both enhance and limit state power and capacities. In situations where state institutions are too weak to act or too reluctant to meet specific concerns, civic associations carry out these activities. In both the United States and Great Britain, retrenchment from central support for local programs was accompanied by encouragement of local public-private partnerships and the blossoming of local development groups structured along corporate lines rather than as traditional voluntary associations. These groups act along with or in lieu of public authority; as such, they constitute a "shadow state" (Wolch, 1990) and exemplify the privatization of public authority discussed below. In contrast, numerous but less visible alternative organizations also emerged to address social needs of marginalized groups, through battered women's centers, homeless shelters, and such; at least initially, many sought to minimize their contact with the state.

But Woods (1992) questions whether this notion of civil society is germane for the African context. The discontinuities in the evolution of civil society in Africa noted by Woods are also features of modern political history in Central Europe and the Caribbean. And in both Africa and Central Europe, democratization pressures involve an organizing principle based on the idea of political accountability and greater differentiation of private and public spheres (Woods, 1992, p. 91). In the African context, an urban middle class whose material interests, social status, and opportunities for political voice are at crisis points pushes for an expanded public arena through increased associational activities; note the profusion of recent groups described by Olowu and Ngau in Chapters 8 and 9, respectively.

It appears that somewhat similar pressures are at play in Central Europe (Starr, 1991), where material interests and associational demands coincide with democratization trends. The abolition of independent civil associations in Central Europe by the Soviet Union meant there was no legal status for such groups but did not eliminate their existence. Such groups were critical means of mobilization in the early stages of democratic transition and important sources of citizen leadership (Weigle & Butterfield, 1992). But the further development of civil society is contingent on a number of contextual factors; as Starr (1991) points out, they can flourish only in the context of a viable, self-regulating market sector and access to capital markets. These features continue to be troublesome for nongovernment associations in Central Europe and Africa. In lieu of this, the links of local groups to a network of international nongovernment organizations have been an important source of expertise and financial resources (van Steenbergen, 1992).

Privatizing local authority. Neither democratization nor restructuring trends necessarily entail a strong public role relative to market forces at the local level. In Chapter 5, Surazska frames the debate over local reforms as a search for ways to replace economic development and social distribution mechanisms at the local level. There are strong pressures for increased local responsibilities for economic activity and social policy across the cases presented here. In each instance, enhanced local development activities are seen as essential for stimulating the growth of the national economy and establishing local revenue capacities. The resources for local policy activism vary by setting; they include privatization of local public assets, joint ventures with private investors, new taxation authority, infrastructure funds, and institutional innovations. But the necessary expertise and opportunities to implement these initiatives are problematic.

Several authors remain skeptical of local governments' capacity to carry out these new roles and responsibilities. To the extent that local governments operate in a setting of "poor capitalism" (Przeworski, 1992), their ability to attract investment and meet citizen demands is doubtful. Péteri notes in Chapter 6, for example, that uneven development patterns and past resource allocation schemes in Hungary restrict the abilities of many cities to compete for new private investment. In a model all too clear in the West, better-off cities will get richer while poor cities get poorer. And to the extent that a new localism generates hypercompetition—Klak and Rulli's term (see Chapter 7)—among communities, past patterns of uneven development will be exacerbated, particularly where horizontal ties and traditions of cooperation among communities are weak, as in the cases described here.

Political fragmentation and limited institutional capacities also proscribe local initiatives; in reaction to past centralization and state planning, many local governments in Central Europe are being carved into multiple units with ambiguous revenue bases and planning authority. In the United States, fragmentation of local development authority stems from past federal programs and continuing efforts to circumvent debt limitations; similarly, the churning of local territorial and organizational structures by central governments in the United Kingdom, Kenya, and Nigeria diminished local capacities.

Political constraints are telling. Newly local budget-making processes reveal preferences—from Hungary to Poland to Nigeria to Kenya—to balance investment with education and cultural needs. Yet local officials are dealing with pressures from central governments—and, indirectly, international organizations—to emphasize economic development agendas and productivity measures. The need to balance these competing agendas of economic management and citizen responsiveness contributes to a political culture and leadership style characterized by Clark and Ferguson (1983) as the "new fiscal populism." Although they focus on the fiscal policy dimensions of this political culture, their description fits the trend toward more activist local public authority evident in the chapters presented here.

The accounts by Chandler (Chapter 3), Jacobs (Chapter 4), and Hula (Chapter 2) depict this as the rediscovery of local activism rather than a novel development. In contrast to previous eras of municipal activism, however, local leaders now move in a decision context where private investors operate at global scales, with resources that dwarf any public share, and at a magnitude and pace that defies local involvement. But, as Chandler puts it, local leaders are asked to perform in this market context as entrepreneurs under public constraints. Many of these constraints are imposed by national governments wary of the fiscal and political consequences of proactive local development efforts. Observers of the new localism in more advanced industrial countries concur on the ensuing power shifts away from elected bodies and popular control and toward those controlling private resources, albeit often through quasi-public agencies. Officials in transitional settings also increasingly seek off-budget institutional arrangements as means of balancing economic and social demands and protecting their budgeting flexibility (Clark, 1992). Given this changed decision context, Chandler maintains that the overall impacts of new local initiatives are less than those of local public entrepreneurs in the late nineteenth century.

Nevertheless, the notion of the local public entrepreneur holds great sway. The interaction of democratization and restructuring pressures at the local level links the legitimacy of local regimes to local economic performance (Fuchs & Koch, 1991) and creates a demand for local public entrepreneurs. Although local entrepreneurial talent is not at issue, the knowledge base and skills to act as public entrepreneurs are evolving slowly and are tied to local contexts in ways that limit their transferability across settings. The decision context in countries experiencing accelerated transitions to democracy is also one of externalized investment power and reluctant devolution. It is further complicated, in Central Europe, by the historically limited role of local authorities in development policies and the all-too-brief period of leadership transition at the local level. In many Central European cities, the negative reaction to planning processes and the lack of local institutional capacity hamper any interest in adopting current Western models of local development initiatives (Enyedi, 1992; Jorgensen, 1992). Furthermore, those previously controlling local governments often were best equipped to administer the operations of democratically elected local governments. Thus, although elected officials have taken on new executive and legislative responsibilities, their autonomy relative to central policy makers is often uncertain and their dependence on local bureaucrats severe. Yet, despite these constraints, Péteri describes Hungarian cities where public entrepreneurs are active (see also Bartkowski, n.d.-a, n.d.-b; Clark, 1992), and Ngau and Olowu provide additional examples from Kenya and Nigeria, respectively.

Missed Opportunities and Preferred Alternatives

Political interpretations of the new localism depend on the implicit logic of the economic, spatial, and political dynamics assumed to contribute to this trend. At a minimum, arguments that a post-Fordist era of flexible production, global cities, and new industrial spaces is emerging suggest a "revaluation of local politics" in these areas (Preteceille, 1990, p. 27). Whether local initiatives can have independent effects on creating or enhancing contextual features encouraging growth is ambiguous. Preteceille (1990) contends that local officials face a hegemonic obligation to promote growth; as Chandler points out in Chapter 3, they may not necessarily think they can have an independent effect on economic forces, but English officials do believe they can conduct certain economic development programs more efficiently than central government can. But even if local initiatives do not have direct effects, Chandler sees activist local governments as potential catalysts for the formation of new growth coalitions supporting new public roles.

In contrast, it is also possible that the new localism masks a depoliticization of the local terrain. Despite the push for local initiatives, local governments labor under central constraints on their policy discretion, their expenditures, their independent voice, and their control of productive assets. In addition, the increased competition for investment isolates communities from each other, although there is some evidence that communities are forming pacts to share certain responsibilities (Clarke & Gaile, 1992). In the accounts presented here, shifts in local opportunity structures appear to skew access to influence to economic interests. Despite the increasingly visible public role, private interests prevail and the actual scope of political choice shrinks. The rise of nongovernment associations, which promises more access but less accountability, may also be a potential means of depoliticizing local government. To the extent that these associations take on tasks performed by neither the state nor the market, they function as a "shadow state" (Wolch, 1990). This was literally true in Poland and somewhat in Hungary, particularly in the activities of the Citizen's Committees and Solidarity; similarly, Olowu's description of the Community Development Associations in Nigeria indicates that they act as a shadow state in rural areas especially.

Although the argument over local autonomy is unlikely to be resolved in a definitive manner and, more important, will continue to lag behind actual practice, the question of alternative paths and models for local practice is compelling. On theoretical grounds, the notion of alternative path choices is important in explaining actual paths of political change (Przeworski, 1986). Theorizing about possibilities requires some sense of counterfactual claims about actual and alternative outcomes; it may include consideration of the effects of missed opportunities and whether historical distance from branching points diminishes the likelihood that previous alternatives are available (Przeworski, 1986, p. 49). Although they do not frame their discussions in these terms, the contributors to this volume trace missed opportunities and preferred alternatives in the reorganization of markets and politics in these diverse local settings.

The language used in this volume is one of possibilities rather than determinants. As Ngau and Olowu document for the African cities, and Péteri and Surazska detail for Central European cities, the possibility of different paths is a continuing matter of local debates. Hula, Chandler, and Jacobs reveal the historical roots of comparable debates on local policy direction in advanced industrial cities. Although not articulated in theoretical terms of political change paths, a substantial literature on variations in policy orientations in American cities (Clarke & Gaile, 1992; Clavel, 1986; Goetz, 1990; Robinson, 1989; Stone & Sanders, 1987) incorporates these concerns with the control and content of local policy choices. Whether this diversity of orientations is indicative of autonomy, however, is a matter of continuing debate (Sassen, 1990).

Perhaps the ambiguity, optimism, and contradictions of new localism trends are best captured by the metaphors used to describe the new local terrain. Terry Nichols Clark, for example, argues that the conflicts between hierarchy and egalitarianism occurring on a global scale generate alternative governance options for local governments in Central and Eastern Europe, each embodying distinct rules of the game. Indeed, Clark (1992) refers to eight "games" that can coexist within societies in an "eight-ring circus": reliance on the traditional Soviet model of centralized planning, unions, clientelism, ethnic/regional identification, the church, the populist leader, voluntary associations, and the new political culture. These ideal types are clearly ordered relative to their distance from strict political regulation of markets; Clark anticipates that, over time, the movement is away from games with rigid, fixed rules toward those allowing more citizen choice or, as Przeworski (1986) would put it, institutionalization of uncertain outcomes. Clark eschews the notion of progression by stages and indicates that simultaneity and overlap will be characteristic features, but it is not clear how one moves from one game to another or, perhaps more accurately, how one game becomes dominant. This raises some interesting questions of norm and rule development and, more broadly, of how alternatives become organized (Przeworski, 1986). To the extent that these games describe actual occurrences, they provide rich description but limited grounds for explaining political change in terms of missed opportunities and preferred alternatives. And in the language used above, these games are neither scale sensitive nor state sensitive, nor are they grounded in theories of economic change.

Nor is Reich's (1991) web metaphor. Reich uses the notion of enterprise webs to distinguish the organizational features of new enterprises from the pyramid structures of another era. In the enterprise web, high-value enterprises consist of working groups and individuals on a global scale linked together by strategic brokers at nodal points; new links are continuously spun out, and each joining place on the web "represents a unique combination of skills." This metaphor reflects the economic and spatial orientations of current theorizing (see, e.g., Amin & Thrift, 1992), but the political implications are limited to an emphasis on human capital investment and are not sensitive to scale effects.

Scott's (1992) emphasis on the need for local institutional infrastructure to minimize transaction costs introduces the notion of political steering. Local officials can steer their localities to avoid premature locks on alternative development paths and to enhance the contextual features amenable to new production systems. This implies more directionality than the basic circus or web metaphor and reintroduces the issue of the conditions under which politics regulates markets at the local level.

Taken together, these metaphors argue for viewing new localism trends in terms of context-structuring processes (Fuchs & Koch, 1991). To the extent that localities are the nexus of democratization and restructuring processes, they will be compelled to create new decision frameworks for the processes of negotiation and bargaining between private interests and public authority. These efforts to create or influence the framework within which groups negotiate characterize the diverse localities described in this volume. Struggles over these context-structuring processes yield the missed opportunities and preferred alternatives to local political change marking the accounts presented here.

Note

1. We are not alone in gratefully acknowledging the vision, energy, and generous assistance of Terry Nichols Clark in supporting comparative urban research endeavors

under the auspices of the Fiscal Austerity and Urban Innovation (FAUI) Project. As president of the International Sociological Association's Research Committee 03 on Community Research, Terry organized panels at the ISA Madrid meetings in 1990 that provided the first opportunity for several of the authors here to learn of each other's work. His encouragement and help with the FAUI data were critical elements in bringing this volume to fruition. We also appreciate the support of the Center for Public Policy Research at the University of Colorado at Boulder; the contributors are well aware that none of the complex international and interpersonal links would have worked without the patient assistance of Luzie Mason at the Center. Mara Sidney and Kathy Wood at the University of Minnesota played similarly important roles in the production of this volume. Finally, Carrie Mullen at Sage Publications stepped in during the final manuscript preparation stages and gently and persistently insisted that we finally make good on our promises and put this into production.

References

Amin, A., & Thrift, N. (1992). Neo-Marshallian nodes in global networks. *International Journal of Urban and Regional Research, 16,* 571-587.

Ash, T. G. (1990). *The uses of adversity: Essays on the fate of Central Europe.* New York: Vintage.

Barnekov, T., Boyle, R., & Rich, D. (1989). *Privatism and urban policy in Britain and the United States.* Oxford: Oxford University Press.

Bartkowski, J. (n.d.-a). *Main features of Polish urban government.* Unpublished manuscript, University of Warsaw, Institute of Sociology.

Bartkowski, J. (n.d.-b). *Polish leaders' attitudes towards privatization.* Unpublished manuscript, University of Warsaw, Institute of Sociology.

Castle, M. (1990). *Transition from a communist regime: The final crisis of the People's Republic of Poland.* Paper presented at the annual meeting of the American Political Science Association, San Francisco.

Clark, G., McKay, J., Missen, G., & Webber, M. (1992). Objections to economic restructuring and the strategies of coercion: An analytical evaluation of policies and practices in Australia and the United States. *Economic Geography, 68,* 43-59.

Clark, T. N. (1992). *Local democracy and innovation in Eastern Europe.* Unpublished manuscript; revised version of a paper presented at the Hungarian Institute for Public Administration, Budapest, March 1991.

Clark, T. N., & Ferguson, L. (1983). *City money.* New York: Columbia University Press.

Clarke, S. E. (Ed.). (1989). *Urban innovation and autonomy: Political implications of policy change.* Newbury Park, CA: Sage.

Clarke, S. E., & Gaile, G. L. (1989). Moving towards entrepreneurial state and local economic development strategies: Opportunities and barriers. *Policy Studies Journal, 17,* 574-598.

Clarke, S. E., & Gaile, G. L. (1992). The next wave: Postfederal local economic development strategies. *Economic Development Quarterly, 6,* 187-198.

Clavel, P. (1986). *The progressive city.* New Brunswick, NJ: Rutgers University Press.

Cox, K. (1992). The politics of globalization: A sceptic's view. *Political Geography, 11,* 427-429.

Eisinger, P. (1973). The conditions of protest behavior in American cities. *American Political Science Review, 67,* 11-28.

Enyedi, G. (1992). Urbanisation in East Central Europe: Social processes and societal responses in the state socialist systems. *Urban Studies, 29,* 869-880.

Fabian, K. (1992). *Privatization and housing policy in Central Europe: Case study of the privatization of the Hungarian housing stock.* Paper presented at the annual meeting of the Association for Public Policy Analysis and Management, Denver.

Frieden, B., & Sagalyn, L. (1989). *Downtown, Inc.* Cambridge: MIT Press.

Fuchs, G., & Koch, A. M. (1991). Corporatism and "political context" in the Federal Republic of Germany. *Environment and Planning C, 9,* 1-14.

Gaspar, J. (1992). Societal response to changes in the production system. *Urban Studies, 29,* 827-837.

Goetz, E. (1990). "Type II policy" and mandated benefits in economic development. *Urban Affairs Quarterly, 26,* 170-190.

Henry, N. (1992). The new industrial spaces: Locational logic of a new production era? *International Journal of Urban and Regional Research, 16,* 375-396.

Holmquist, F., & Ford, M. (1992). Kenya: Slouching toward democracy. *Africa Today, 39,* 97-111.

Jorgensen, I. (1992). Urban planning and environmental policy in the context of political and economic changes in Central Europe: International seminar, Central European University, Prague, January 1992. *International Journal of Urban and Regional Research, 16,* 649-650.

Judd, D., & Parkinson, M. (Eds.). (1989). *Regenerating the cities.* New York: Scott, Foresman.

Keeler, J. T. S. (1993). Introduction: New perspectives on democratic reform. *Comparative Political Studies, 25,* 427-432.

King, D. S., & Pierre, J. (Eds.). (1990). *Challenges to local government.* London: Sage.

Logan, J. R., & Molotch, H. (1987). *Urban fortunes: The political economy of place.* Berkeley: University of California Press.

Logan, J. R., & Swanstrom, T. (Eds.). (1990). *Beyond the city limits: Urban policy and economic restructuring in comparative perspective.* Philadelphia: Temple University Press.

Mabogunje, A. L. (1990). Urban planning and the post-colonial state in Africa: A research overview. *African Studies Review, 33,* 121-203.

Mayer, M. (1989, September). *Local politics: From administration to management.* Paper presented at the Cardiff Symposium on Regulation, Innovation and Spatial Development, University of Wales.

Piore, M., & Sabel, C. F. (1984). *The second industrial divide.* New York: Basic Books.

Preteceille, E. (1990). Political paradoxes of urban restructuring: Globalization of the economy and localization of politics? In J. R. Logan & T. Swanstrom (Eds.), *Beyond the city limits: Urban policy and economic restructuring in comparative perspective* (pp. 27-59). Philadelphia: Temple University Press.

Przeworski, A. (1986). Some problems in the study of the transition to democracy. In G. O'Donnell, P. Schmitter, & L. Whitehead (Eds.), *Transitions from authoritarian rule: Comparative perspectives* (pp. 47-63). Baltimore: Johns Hopkins University Press.

Przeworski, A. (1992). The "East" becomes the "South"? The "autumn of the people" and the future of Eastern Europe. *PS, 24,* 20-24.

Reich, R. (1991). *The work of nations.* New York: Knopf.

Robinson, C. J. (1989, Summer). Municipal approaches to economic development: Growth and distribution policy. *Journal of the American Planning Association,* pp. 283-295.

Sassen, S. (1990). Beyond the city limits: A commentary. In J. R. Logan & T. Swanstrom (Eds.), *Beyond the city limits: Urban policy and economic restructuring in comparative perspective* (pp. 237-242). Philadelphia: Temple University Press.

Savitch, H. V. (1988). *Post-industrial cities.* Princeton, NJ: Princeton University Press.

Scott, A. J. (1992). The collective order of flexible production agglomerations: Lessons for local economic development policy and strategic choice. *Economic Geography, 68,* 219-232.

Starr, F. (1991). The third sector in the Second World. *World Development, 19,* 65-71.

Stohr, W. B. (1990). *Global challenge and local response: Initiatives for economic regeneration in contemporary Europe.* London: Mansell.

Stone, C., & Sanders, H. T. (Eds.). (1987). *The politics of urban development.* Lawrence: University Press of Kansas.

Tarrow, S. (1991). Aiming at a moving target: Social science and the recent rebellions in Eastern Europe. *PS, 24,* 12-19.

van Steenbergen, B. (1992). Transition from authoritarian/totalitarian systems. *Futures, 24,* 158-166.

Walton, J. (1990). Theoretical methods in comparative urban politics. In J. R. Logan & T. Swanstrom (Eds.), *Beyond the city limits: Urban policy and economic restructuring in comparative perspective* (pp. 243-257). Philadelphia: Temple University Press.

Weigle, M., & Butterfield, J. (1992). Civil society in reforming communist regimes: The logic of emergence. *Comparative Politics, 25,* 1-23.

Wolch, J. R. (1990). *The shadow state: Government and voluntary sector in transition.* New York: Foundation Center.

Woods, D. (1992). Civil society in Europe and Africa: Limiting state power through the public sphere. *African Studies Review, 35,* 77-100.

2

The State Reassessed

The Privatization of Local Politics

RICHARD C. HULA

It is probably not an overstatement to claim that the past 10 years have seen a global reassessment of the role of government. Against all reasonable expectations, the decade began with the successful efforts of the Thatcher government to dismantle key elements of the British welfare state. Similar antistate themes were taken up by a number of conservative governments in Western Europe and, of course, in the United States by the Reagan/Bush administrations. To the continued surprise of many, a number of Third World governments also sought to invigorate their often stagnant private sectors. Noncommunist governments throughout Eastern Europe hope a revived private sector will promote economic growth and political stability. Finally, most of the states emerging from the disintegration of the Soviet Union see a healthy and powerful private sector as essential for future economic and political development.

Certainly in the United States the decade of the 1980s was a difficult time for those interested in increasing the scope of government. Rather than expansion, a more popular rallying cry has been for government at all levels to do and spend less. A wide range of political leaders, including many long sympathetic to an activist model of government, have engaged in a general reassessment of the appropriate scope and definition of public goals, and the strategies by which these goals

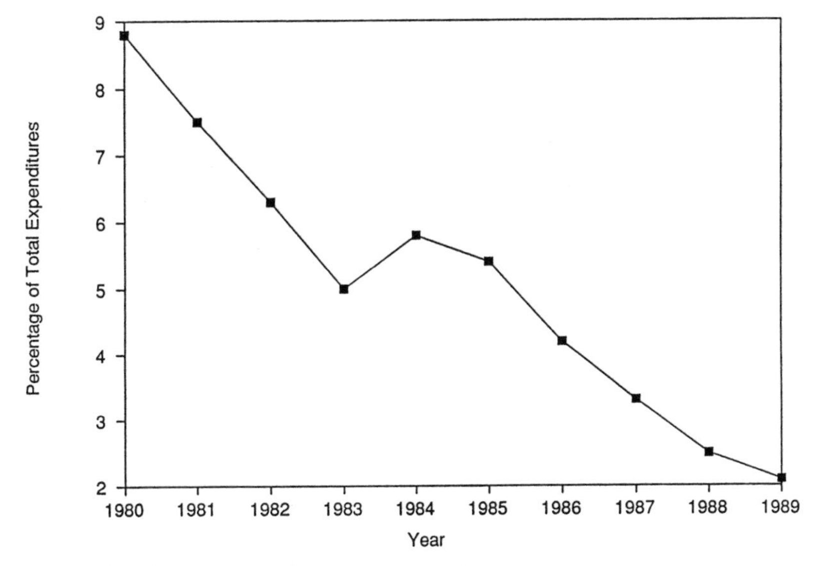

Figure 2.1. Federal Contribution to County Budgets
SOURCE: Data from Pammer (1992).

should be sought. This review has taken place within the context of widely held views that government is overextended and that some measure of retrenchment is in order. Although it is possible to cite modest fiscal retrenchments at the federal level, it is at the local level that one finds truly significant structural change. Such change includes both a shifting pattern of intergovernment relations and a modification of key actors central to the local political process itself. This chapter argues that these changes have had a significant impact, not only on the form of local politics but also on its substance.

Federal and Local Restructuring Trends

The 1980s saw a reversal in the long-term growth of direct federal involvement in local policy arenas. The most obvious element of this retrenchment consisted of reductions in federal expenditures targeted either directly to local governments or to programs designed to address policy issues of particular interest to local governments.

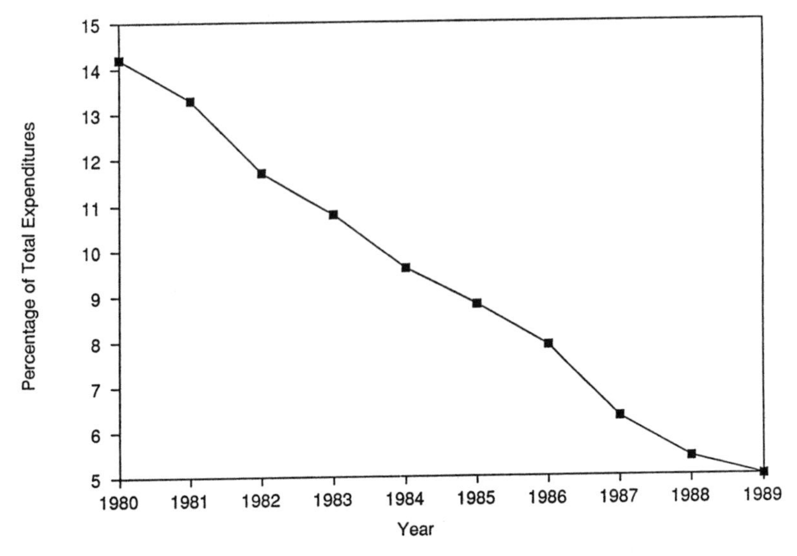

Figure 2.2. Federal Contribution to City Budgets
SOURCE: Data from Pammer (1992).

Federal Restructuring Trends

Figures 2.1 and 2.2 document the broad fiscal impact of this federal policy shift (for a more complete breakdown, see Pammer, 1992). Figure 2.1 reports the federal percentage of county general revenue. In 1980 this proportion stood at approximately 9%; by 1989 it dropped to around 2%. Figure 2.2 shows a similar decline for cities, with federal contributions declining from approximately 14% in 1980 to about 5% in 1989. These data tend to underestimate the impact of federal cutbacks, because the reductions were focused in specific programmatic areas. Federal support for employment programs, community development subsidies, and low- to moderate-income housing efforts was subject to particularly severe reductions.[1]

Although expenditure data help illuminate federal policy shifts, they capture only one important element of the story. Also important are federal efforts to shift the implementation of key elements of social policy to the private sector. This includes, for example, an increased reliance on voluntary associations, nonprofit organizations, and, most important, markets to produce and distribute collective goods. Critics

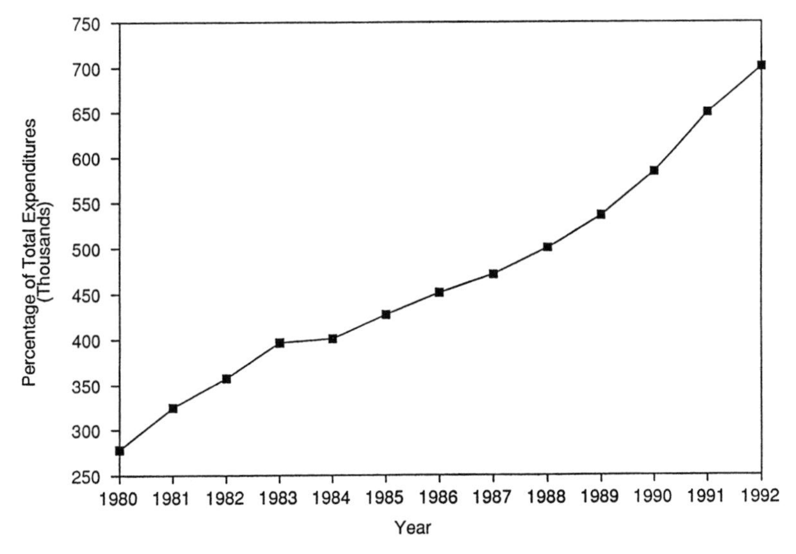

Figure 2.3. Federal Outlays to Individuals

SOURCE: Data from Pammer (1992).

sometimes discount these nonfiscal elements of federal policy as mere window dressing to mask the impact of budget cutbacks. Although there is certainly some truth to this charge, Figure 2.3 shows that even as direct federal support of county and city spending has declined, overall federal outlays for individuals have continued to climb. This suggests that federal policy has shifted in how specific policy is implemented as well as the level at which it is supported. Although the evidence of structural change is largely anecdotal, numerous commentators claim that such a process is occurring (Osborne & Gaebler, 1992).

Local Restructuring Trends

Pressure to engage the private sector in the design and implementation of public policy is even greater at the local level. Sometimes these efforts are mandated by other levels of government. For example, federally funded employment programs must be designed and implemented by representatives of the business community in Private Industry Councils. Urban Development Action Grants require a showing of substantial private investment. Current Housing and Urban Development programs often require substantial tenant participation. Other pressures to seek private investment include severe revenue constraints, deteriorating economic conditions, and an increasingly hostile and alienated citizenry (MacManus, 1992). There exists a bewildering array of strategies to increase the role of the private sector in the local political system. Some examples seem to be consistent with long-standing demands that government activities be privatized (for a discussion of the specific forms privatization can take, see Savas, 1987). Prominent examples include governments entering formal contractual relationships with private service providers and direct consumer subsidies through voucher programs. Other examples are much less direct and more complicated. Typically, they involve elaborate public-private associations in which significant public resources and decision-making authority are delegated. However, the conceptual underpinnings driving efforts to engage the private sector in the policy process are remarkably consistent, centering on two related arguments. The first claims that markets are almost always able to produce and distribute goods and services more efficiently than can the public bureaucracy. In addition, resource decisions are seen as more rational when left to individuals who choose alternatives on the basis of their own preferences through a market bidding process. An infusion of market (or at least marketlike) discipline is seen as having the particularly important benefit of controlling a widely predicted tendency of citizens to overconsume publicly provided goods and services.[2]

Figure 2.4 outlines a set of dimensions that summarize some important attributes of specific local efforts by local government to harness the power of the private sector. Three dimensions are identified: clarity of goals, separation of public and private actors, and level of public oversight. Clarity of goals centers on program intent. Goals can range from very concrete and specific, such as refuse collection, to relatively complex and undefined, such as economic development. Three specific elements of *goal identification* are listed: whether the program/policy has a single or multiple goals, whether goals are clearly articulated, and whether program outcomes can be measured in some standard unit. Although this unit is most often monetary, there is no logical reason others might not be used. *Role separation* refers to how clearly one can identify public and private roles within a policy field. *Level of public oversight* refers to the operational control held by public officials. Taken together, these three dimensions suggest a matrix that can order very diverse forms of privatization.

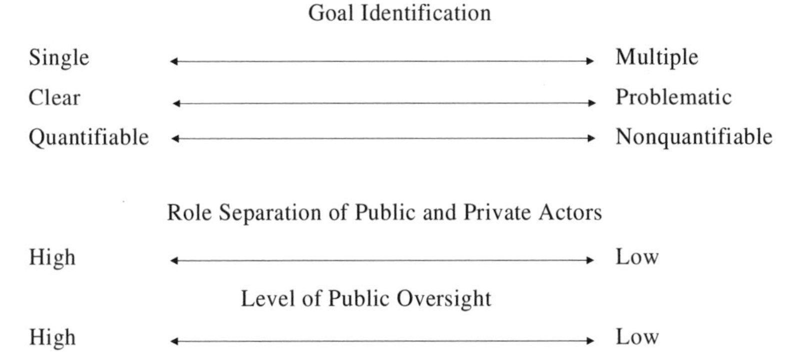

Figure 2.4. Critical Dimensions of Privatization

Effects of Privatization Trends

There exists an extensive empirical literature on efforts by all levels of government to use the private sector to produce and deliver goods and services. For the most part, however, this work takes a rather narrow view of privatization in which goals are well defined, with high institutional role separation and public oversight. Privatization is most often conceptualized as engaging the private sector to deliver a set of specific goods and/or services. This focus has produced an empirical literature that reveals a good deal about the relative economic efficiency of alternative implementation strategies. This is particularly true for contracting (for an excellent review of these issues, see Donahue, 1989). Obviously, measurement issues are central to such evaluation efforts. How can one know the actual benefits and costs of policies? Over what period of time must they be observed?[3]

Although these economic issues are of interest and are certainly important, they tend to ignore a number of broader theoretical and practical concerns. Privatization not only affects the costs of distributing social goods, but may have important effects both on patterns of political participation and the process by which social policy is created. This is likely to be true particularly in some of the more complex forms of privatization, where the specific output is less clearly defined and public and private actors are less clearly separated. This argument rests on three general propositions, which are discussed in turn below.

The implementation of public policy by private sector institutions will likely have a direct effect on the substantive goals and outcomes of that policy. There is broad consensus that policy design, creation, and implementation are hopelessly intertwined. Evidence for this view can be obtained from a wide variety of sources. At the policy-creation level it is well recognized that administrative staffs play an important role (Fesler & Kettl, 1991). Pressman and Wildavsky's (1984) well-known analysis of the Economic Development Administration in Oakland produced a score of implementation studies showing how the substantive impacts of programs depend on the actors central to the implementation process. Work pioneered by Michael Lipsky (1980) revealed that in some policy areas significant discretion occurs at relatively low levels of the bureaucracy. In short, the empirical evidence simply overwhelms traditional claims of public administration and bureaucracy that one can functionally separate such policy roles. Given such evidence, some see efforts to describe a policy process in which implementation can be separated from policy development as little more than a simpleminded straw man argument used to promote a particular ideological view. However, to dismiss the policy/implementation dichotomy as a poorly drawn straw man is to miss the important fact that many who actually create policy still argue that such a separation is possible and continue to act on that belief. This assumption is evident in political discourse on the relative merits of privatization. Both popular and academic debates largely focus on the relative efficiency of alternative implementation strategies. It is assumed (often implicitly) that the process of setting policy goals will continue to reside with public decision makers, presumably elected officials. If privatization forces control from the public sector, it is essential to identify where decision-making authority resides and what effects these shifts have on current and future policy.

Privatization is not inevitably associated with a reduction in the overall scope and activity of government. Those supporting increased levels of privatization often present it as a technique to reduce or at least contain the size and scope of government. Such a claim might initially seem quite reasonable, because most privatization schemes require fewer public employees than would be needed if the good or service were provided directly by government bureaucracy.[4] It is equally clear, however, that only in the case of total government exit is a reduction in overall cost a logical consequence of privatization. Even here the argument is severely limited, because government exit in a particular substantive area might generate large public costs in other sectors. The association between functional scope of government and degree of privatization would seem certainly an empirical rather than a theoretical issue. Once again, as one moves from less to more complex forms of privatization, it is less clear what the expected outcomes should be. It is possible, for example, that those interested in expanding the role of government might also use privatization as an implementation strategy.

The efforts of local government to promote economic development show how complex the association between government scope and privatization really is. Certainly all large cities, as well as many smaller cities, are currently investing public resources toward the goal of generating local economic development. Although there is an enormous variation in strategies implemented by specific cities, they are very often implemented and sometimes developed in the private sector. Through a variety of fiscal conventions, many of these efforts occur off-budget. Specific projects are implemented by private actors, with little if any public review. Such projects may have the effect of expanding rather than reducing the functional scope of local government. Local governments may even find themselves in completely unfamiliar functional areas, such as equity partners in commercial operations (see the discussion offered by Eisinger, 1988).

Privatization is related to future policy development (development in this case is to be understood in terms of both process and substantive outcome). The initial assumption presented above suggests that privatization is likely to have an independent effect on program outcomes. Beyond this short-term effect, privatization is likely to have longer-term effects, both on the process by which future policy is created and on the substantive outcomes of that process. The theoretical justification of this assumption is derived from recent work in urban political economy that has reconsidered fundamental issues of governance in American cities. Much of this debate has been stimulated by Paul Peterson's (1981) claim that structural constraints imposed by the economic system and the U.S. federal political system severely constrain the ability of local governments to set policy. Peterson argues that local governments are particularly constrained in their ability to implement redistributive policies likely to be opposed by mobile capital interests. Rather, local governments seek out development policies that serve to increase total local capital. He claims that such policies represent, from the perspective of local policy makers, the maximization of public interest. Although there is little serious debate concerning the fundamental assertion that such economic and political constraints exist, two streams of research suggest important qualifications to Peterson's thesis. These are current work on local political regimes and local political culture.

Regime theory directly reasserts the importance of politics in an understanding of local political economy. Stone's (1989) work on Atlanta shows that local politics remains a process of making choices:

> Choice enters into the creation of a set of arrangements whereby accommodation is reached between the wielders of state power and the wielders of market power. Put another way, local politics matters, but it is shaped by the political-economy context. (p. 5)

Elkin has also stressed the ability of political decision makers, if not to overcome structural constraints, at least to exercise some measure of policy autonomy:

> State officials are thought to have choices, and these choices are critical to whether the political economy flourishes, remains stable, declines or crashes. But state officials don't operate in an empty world. The world is structured; it presents obstacles and opportunities, both of which may be promptly ignored since these structural arrangements require interpretation. (Benjamin & Elkin, 1985, p. 7)

A good deal of case study material gives strong intuitive support to the view that leadership and politics can have a significant impact on the substance of local policy.[5]

Most efforts to link the form of the political regime with policy outcomes have consisted of case studies describing the formation of local coalitions and the public decisions taken by these coalitions. The issue of economic development is by far the most thoroughly examined issue, although a limited number of scholars have examined a wider range of issues in specific cities.[6] This analysis argues that policy content is a function of the interests and the capacity of the governing coalition. As noted by critics, this formulation suffers from a lack of specificity and provides little predictive power. Nevertheless, it does offer a broad theoretical argument linking implementation strategies and access to the political decision-making process. Such access is an essential element of membership in any governing coalition. Thus any implementation strategy that somehow modifies access to the decision-making process would likely have an influence on the makeup of the governing coalition. To the extent that one can predict policy outcomes to regime makeup, there should be a direct tie between implementation strategies and policy.

Stone (1987) suggests that both regime structure and policy outcomes across U.S. cities are consistent enough to identify three general types:

1. The *corporate regime* is dominated by local, usually central-city, economic elites. The corporate regime typically seeks an active, downtown-centered development policy.
2. The *progressive regime,* uncommon in the United States, is dominated by lower- and middle-class neighborhood interests. Such a regime is relatively more interested in redistributive policy.
3. The *caretaker regime* is dominated by small business owners and homeowners. Its overriding policy goal is to minimize the size of the public sector (see also Whelan & Young, 1989).

Such efforts to identify varieties of local regimes suggest a second important challenge to Peterson, the notion of local political culture. Clark and his associates begin with the assumption that not all local governments respond to political issues in the same way.[7] This variation is in part a function of a local political culture that sets broad parameters on local decision making. Obviously, this notion of political culture has substantive significance only if local authorities have the ability to make choices.

Clark describes three traditional local political cultures in the United States. Ethnic culture organizes local politics on the basis of ethnicity and mutual reciprocity. Democratic and Republican cultures are organized around party ideology. Clark argues that a fourth culture, "new fiscal populism," is increasingly challenging the traditional cultures. New fiscal populism is seen as combining elements of traditional left social and right fiscal ideology. For example, these new leaders present a liberal social agenda while at the same time promoting a conservative fiscal agenda. Clark argues that this hybrid political culture helps explain what might seem to many to be inconsistent political behavior on the part of local leaders. The point is that local decision makers are not acting irrationally, or even simply opportunistically. Rather, such decision makers are applying a new logic to political issues. Privatization is seen as an important element of this emerging political culture. What makes all of this particularly interesting is the assumption that political culture has a direct link to local policy outcomes.[8]

In sum, regime theory attempts to describe the process by which governing coalitions are formed among the set of possible interests within a community. Those interested in local political culture tend to focus more directly on issues of norms and ideology that are thought to drive the policy process. Both perspectives, however, discuss direct policy outcomes only in the most general of terms. The long-term effects of specific political outcomes on future levels of political participation and policy are less often considered. The extent to which privatization might actually cause political changes as well as serve as a means of implementing current policy is a question that deserves further analysis.

A Case Study in Local Restructuring

A key theme of this chapter is that privatization is a more complex phenomenon than is often assumed. It is not simply an alternative implementation strategy, but a broad political strategy that may transform important elements of a local political process, including both policy outcomes and citizen participation. To illustrate the general argument, I present a brief case study of local economic development in Baltimore, Maryland. The case is particularly interesting given the importance of private actors in the renewal process and the widely held view that the city's effort has been quite successful. Obviously, no single case offers statistical evidence of a general phenomenon. It can, however, provide a rich context in which important clues to understanding social and political change can emerge.

Local Economic Development: The Case of Baltimore

A very substantial case study literature has emerged that describes the efforts of cities to promote local economic growth. Although cities have long sought development, such efforts have been greatly accelerated in the past decade in the face of continued internal economic decline and sharp reductions in federal urban programs. Not unexpectedly, there is a good deal of variation in strategies adopted by specific cities. However, in almost all cities the private sector has been seen as a primary (and often the only) source of investment.

Like most large U.S. cities, Baltimore's major redevelopment initiatives have focused on downtown renewal (for a more complete discussion, see Hula, 1990). Even a brief review of these efforts reveals the apparent paradox of the city's political leadership's simultaneously seeking to reduce the scale of government while embarking on a series of economic development initiatives.[9] This paradox is resolved by analysis of the privatization strategies used. Although these include traditional forms of privatization, the city has relied most heavily on the quasi-private corporation to implement redevelopment policy.[10] For example, the Central City Development Corporation currently implements much of Baltimore's downtown renewal program. Major industrial development efforts have been channeled through the Baltimore Economic Development Corporation (Berkowitz, 1984, 1987). Other, more specialized organizations have also been created. Some of these deal with specific tasks, such as the management of the city's convention center and Hyatt Hotel, the National Aquarium, and the municipal marina. Others have been given broader charges.

The private sector has played an important role in the Baltimore renewal effort since its inception. Indeed, a good deal of the strategic planning eventually incorporated into public plans actually took place in the private sector. This process was begun in the early 1950s, when the Greater Baltimore Committee and the Committee for Downtown created a private planning unit to review alternative redevelopment strategies.[11] What emerged was the Charles Center redevelopment plan (Bonnell, 1982). It identified a 33-acre core in Baltimore's central business district for redevelopment. The $180 million project was to transform the area into a center of modern offices, retail establishments, and some apartments. The focused nature of the plan was justified based on the notion that for downtown redevelopment to occur there needed to be a single dramatic focal point to show that such development was possible.

Two critical assumptions were built into the Charles Center plan. The first identified downtown commercial interests as the key to economic revitalization. Such investment was seen as a means to integrate Baltimore's local economy into an emerging national economy based not on manufacturing but on information and services. Such an integration was seen as having very broad distributional benefits for the entire city. A second assumption involved the critical role of private capital. In fact, the project was conceptualized largely as a private venture to be facilitated by the local government, primarily through land acquisition and clearance. Of the estimated $180 million cost of the project, only $8.3 million was to come from public funds.

The Charles Center redevelopment plan was submitted to the city in 1958. It was received with enthusiasm and quickly replaced the city's own renewal design. Lyall (1980) suggests that the city plan suffered from two major faults: First, it failed to gain the support of the business elite because it largely ignored the downtown; second, it was greeted with hostility in a number of the city's neighborhoods that had been targeted for freeway construction.

The complex network of institutions that has evolved around the Charles Center plan and its successors is sometimes referred to as Baltimore's "shadow government" (Stoker, 1987). The most prominent organization in the shadow government has been the agency charged with implementing central-city renewal. There have been three such agencies: the Charles Center Management Office, the Charles Center Inner Harbor Management Corporation, and the Center City Development Corporation. The formal political status of these corporations is difficult to specify. All have received significant public operating funds to design and implement redevelopment policy. In addition, all have had a good deal of influence on capital spending in the downtown area.[12] From this perspective they seem clearly to be public agencies. Nevertheless, corporation officials have consistently denied public access to records and meetings on the grounds that they are private organizations operating under contract with the city. In making this claim, the corporations not only secure privacy for their activities but avoid a number of constraints facing public officials, ranging from mandatory relocation payments for displaced individuals to competitive bidding.[13]

A number of reasons have been put forward to support the creation and operation of the shadow government. The most common is a desire to identify a nonbureaucratic alternative to city government. Removing day-to-day redevelopment decisions from the public domain is seen as both reducing the influence of politics and providing the environment needed to ensure private sector participation. City agencies claim a need to operate in relative secrecy to protect sensitive negotiations with developers. Finally, multiple and sometimes overlapping organizations are seen as a means of promoting interagency competition for resources. Such competition is seen as a way to promote maximum effort and efficiency in program implementation.

Many observers see the Baltimore redevelopment program as a model for other cities. Writing in *Fortune* magazine, Gurney Breckenfeld (1977) presents the popular view:

> By almost every standard measure of trouble, Baltimore should be firmly trapped in the vortex of urban decay. It is an old, conservative, blue-collar industry place. . . . Yet Baltimore, which H. L. Mencken called the "ruins of a once great medieval city," is making an extraordinary comeback, both physically and—more important—psychologically. (p. 196)

For Breckenfeld the source of Baltimore's success is clear:

> The critical ingredient of that revival is two decades of intelligent teamwork between local officials and private business leaders. Their strategy, established at the outset, has been to convert the heart of the city into a culturally rich, architecturally exciting magnet where both affluent and middle class families will choose to work, shop and live. (p. 196)

Certainly city officials can point to convincing evidence of a very widespread central-city physical renewal, much of it the result of private investment. Moreover, the city has, in defiance of conventional wisdom, transformed itself into a major Maryland tourist center.

Public debate about the distributive impact of redevelopment was stimulated by the 1986 publication of *Baltimore 2000* (Szanton, 1986). This report, prepared for the Morris Goldseeker Foundation, offered a review of the current status and likely future of the city. It took a decidedly pessimistic view of current trends, arguing that the immediate future of Baltimore was one of decline, not renaissance. This process of decline was seen as a pattern of uneven development that would, in effect, create two Baltimores:

> Baltimore may complete a pattern, already visible, of a "double-doughnut" of concentric rings. The center would contain a business, cultural and entertainment center that remained strong because it served the whole metropolitan area, and attractive housing for the well-to-do. The center would be ringed by the decaying and much more populous neighborhoods of the poor and dependent, very largely black. These in turn would be surrounded by middle- and upper-income suburbs, very largely white. (Szanton, 1986, p. 21)

Baltimore 2000 denied any automatic link between downtown prosperity and neighborhood well-being. This conclusion is consistent with other work suggesting that even where downtown renewal did create jobs, they were relatively low paying and had little potential for advancement. The limited number of more attractive jobs often went to non-city residents.[14] The report called for a dramatic shift of priorities to neighborhood and community issues, stressing the importance of infrastructure investment, particularly in education.

Some of the debate over the distributive effects of redevelopment could be resolved by better data on program impact, particularly job creation. Other issues need to be resolved at a more abstract, ideological level. One key issue centers on the political process that has created and implemented the redevelopment effort. There is certainly room for genuine debate over the independence of Baltimore's network of redevelopment agencies and corporations, but it is difficult to dispute that a great number of critical economic decisions have been made by private organizations with only very limited public input. To be sure, those active in the process saw themselves as working unselfishly in the interests of the city. The underlying assumption that development was in the interest of the entire city was never seriously questioned. This sentiment is captured in a comment by Frank Knott, who, as a developer and a member of the city's Industrial Development Board, was asked whether he might not face a conflict of interest: "I think conflict of interest is the wrong term. Conflict of interest is something that hurts someone. Our involvement is based on the common good, also recognizing that our involvement will benefit us" (quoted in Smith, 1980). This vision of the public good was typically coupled with a mistrust of politics and an outright hostility to public bureaucracies. Commenting on the value of strong executive leadership, a publication of the Historic Baltimore Society put it quite frankly:

> One hopes that there will come a day in the city's future when it can afford the slow erratic process of local municipal politics—the luxury of hesitant and compromising democracy. However, during the last 30 years the city's problems have been too large, and the inherent divisions in the city too deep, to allow it to move forward without a person of power, determination, and vision in the mayor's office. (Arnold, 1982, p. 247)

It is clear that in addition to the issues of substantive benefits of redevelopment, Baltimore poses fundamental questions about the efficacy of local politics. Those stressing efficiency and "getting things done" often see the local political system as simply a set of hurdles to overcome. Values of representation and accountability are replaced by an emphasis on concrete results.

Critics have made the explicit argument that the city's development policies have had the effect of significantly reducing the autonomy of local political officials. If one is committed to private sector investment, one is inevitably constrained by private sector values. Consider the reaction of the Baltimore City Council to two recent requests by developers. The first was a proposal by the IBM Corporation to add an additional 26 stories to its Inner Harbor office building. The second was a request by the Rouse Corporation to raze what was widely regarded as a historical building and operate a temporary parking lot beyond the normal period allowed for such facilities by the city. Both requests were in clear violation of city policy, but both were reluctantly approved. The majority of the council cited a continuing need to be "responsive" to private developer needs. Even with a history of significant redevelopment and an apparent high demand for additional development opportunities, city officials felt unable to impose even such modest restrictions on private investment.[15]

It should be stressed, however, that 20 years of central-city renewal activity in Baltimore does little to deny the importance of political leadership. In fact, quite the opposite is true. There is no serious debate among either local activists or scholars that Mayor Donald Schaefer played an absolutely central role in creating and empowering the Baltimore growth coalition. As mayor from 1971 to 1987, Schaefer was tireless in his efforts to engage private resources for his vision of a renewed Baltimore. The issue is not whether Schaefer was in some sense captured by private interests, but rather the range of policy goals a mayor may seek to implement through such a private sector strategy. Schaefer articulated a downtown redevelopment agenda and was quite successful in building a coalition to promote those goals. The potential to redefine goals of this dominant coalition has been directly addressed by Donald Schaefer's successor, Kurt Schmoke. Elected in 1987, the city's first African American mayor has sought to deemphasize physical renewal in favor of a broad effort to stimulate a renewal of the city's education system. To implement this policy, Schmoke has sought to broaden the governing coalition of the city. In particular, he has worked closely with a community organization called Baltimoreans United in Leadership Development (BUILD). BUILD has a solid base of support in Baltimore's African American churches, and has come to serve as something of a counterweight to the Greater Baltimore Committee in the policy process. Yet two central features of the Schaefer regime remain. First, a strong emphasis continues on private policy development and implementation. Moreover, Schmoke has worked very hard to maintain key elements of the growth coalition by stressing the direct economic gains of his "human capital" program.

Conclusions

It is, of course, impossible to generalize from a single case. However, the political transformation observed in Baltimore does provide an interesting model through which we might understand local responses to the changing political and economic environment faced by cities in the United States. This model can be translated into the set of propositions presented below, to be tested in future empirical work.

Local governments are restructuring in ways that mobilize types and levels of private resources not normally available to purely public institutions. There can be little doubt that significant private resources were generated and expended in an effort to reach organizational goals in the Baltimore redevelopment effort. The important point about this investment is that it was generated not by government coercion or appeals to civic virtue, but through private analysis of likely economic return. Note that these resources were not purely financial; private leadership skills were also essential to the effort. The case study presented above supports the common view that private organizations may have a wider range of incentives to use in generating desired outcomes.

Local governments are restructuring in ways that shift program goals toward traditional economic elites.[16] There is a good deal of intuitive evidence for this proposition in the literature describing central-city renewal efforts. Certainly the Baltimore case provides evidence of significant goal displacement. Although the renewal in Baltimore is tied to a powerful symbolic commitment to the entire city, the evidence of widely shared distributive benefits is meager. Rather, the redevelopment has largely benefited downtown property interests.[17] What is not clear from the anecdotal literature, however, is whether other political agendas can be supported through a similar process of local restructuring. The current administration in Baltimore provides an example of an effort to create a governing coalition that embraces a number of goals beyond those typically identified by the economic elite.

Local governments are restructuring in ways that may reduce popular control. Almost by definition, privatization creates an additional barrier between those who implement policy and the public at large. This is surely true in the Baltimore case. This barrier can serve to reduce popular control in a variety of ways. At a minimum, it introduces an additional information demand on citizens. That is, one must be able to identify appropriate actors within a policy arena. Sometimes these information costs can be quite high, because private organizations often attempt to shield themselves from public scrutiny. Again, the Baltimore redevelopment case is instructive on this point.

Sullivan (1987) has argued that privatization may have even more profound effects on policy making and implementation. Based on a review of a wide variety of judicial rulings that have held private organizations are not bound to honor the range of constitutional protections that routinely constrain government actions, he concludes that widespread privatization may actually have the effect of reducing citizens' constitutional rights. The case of Baltimore gives some modest support for this view in that citizens did lose access to once-public information as well as concrete benefits such as displacement compensation.

There is, however, an important qualification to the view that privatization inevitably reduces public control. From the perspective of a group newly incorporated into the political process, quite the opposite will seem to be true. If, for example, resident managers are more accessible than are conventional managers, privatization may be empowering to public housing residents even as it reduces overall popular control of public housing. Current redevelopment plans in Baltimore seek to involve community-based activists as well as economic elites. Thus a key normative issue surrounding privatization is whether it can be implemented to include nontraditional elites.

Embedded in these findings is an outline of a future research agenda. Privatization needs to be explored on three separate levels. First, do concrete efforts succeed in meeting specific programmatic goals? This is, of course, the level at which the extant literature on privatization focuses. In fact, a good deal is known about the direct effects of simpler forms of privatization. For example, in an excellent review of this literature, Donahue (1989) makes a very convincing case as to when contracting will achieve the generally desired goal of reducing costs of public authorities. Even at this level, however, a good deal less is understood about the impacts of the more complex forms of privatization discussed in this chapter.[18]

Even for relatively simple forms of privatization, a second level of inquiry is important. This level focuses on possible externalities. That is, what unanticipated consequences (desirable or undesirable) might such efforts create? For example, cost reductions associated with contracting often result in lost wages and benefits for workers. Tenant managers have been alleged to discriminate against the very poor in public housing. To be sure, policy externalities are not always negative. However, it is essential that such effects be considered in any evaluation of privatization.

A final level of analysis turns to the general conceptual issues raised by the propositions derived from the case study of Baltimore. That is, how might privatization transform the political process? The question is essentially whether politics and markets actually serve to manage similar social, political, and economic demands. A lack of equivalence is suggested by anecdotal evidence that privatization may induce a restructuring of public policy. This restructuring can occur in two ways:

1. Privatization may shift political opportunity by changing the mix of group interests able to influence political decision makers. Put simply, privatization may affect the rules of the political game in ways that benefit some and disadvantage others.
2. Because privatization is rooted in a market-defined notion of self-interest, it may have the effect of crowding out other expressions of preference derived from altruism or collective interest.

If a strong public commitment to privatization carries with it a concurrent commitment to economic logic, a number of important constraints to public decision making emerge. Of particular importance is the central role given to structural characteristics of specific goods and services in determining whether such goods should be privately or publicly produced and distributed, and the dominance assumed by individual interest in making public choices. The net effect is to reduce policy options of decision makers. A decision rule based on structural criteria directly limits the universe of goods and services appropriate for public production and distribution to technical public goods (for an application of this logic, see Savas, 1987). Similar policy constraints are introduced by focusing on individual interests as a primary choice criterion in selecting policy. Here policy options are reduced by minimizing the relevance of collective choice mechanisms. This point is nicely made by Jane Mansbridge (1990):

> Designing institutions when participants have both selfish and nonselfish motives requires attention to variations in context, individuals and the way individuals learn from institutions. The seemingly cautious strategy of designing institutions to work only on self interest, so that if there is little or no public spirit the institutions will work anyway, will in some conditions erode whatever public spirit might exist. (p. xxi)

The question of how we ought to design political institutions raises fundamental issues about the appropriate role of the political system in society. Many proponents of privatization identify this role as simply one of producing and distributing a set of goods and services at the lowest possible cost. Hero (1987) has criticized efforts to evaluate local government solely in terms of the empirical distribution of concrete goods and services. He argues that to stress a politics of consumption is to ignore other aspects of the political system, such as participation. It is participation that "might broaden citizens beyond their narrow interests as consumers" (p. 674). Hero cites with approval Bachrach's discussion of democratic elitism, in which he argues that contemporary defenses of democratic theory often are based on the belief that

> the value of ordinary individuals should be measured by the degree to which "outputs" of the system, in the form of security, services, and material support benefit them. On the basis of this reasoning, the less the individual has to participate in politics on the input side, the better off he is. The democratic elitist implicitly rejects the contention of classical theorists that interests also include the opportunity for development which accrue from participation in meaningful decisions. (p. 674)

Thus, if one reconceptualizes the role of government from the specific task of producing and distributing concrete goods to the broader task of implementing social and political goals, the functional equivalence of markets and politics begins to break down. With this collapse comes the need to reassess much of the intellectual basis for privatization. Perhaps valid interests exist at levels of social organization other than the individual that are not well served by private markets (for more complete discussion of this issue, see Elkin, 1987; Stone, 1985).

The key issue facing those interested in privatization is to identify those circumstances where privatization can be implemented without threatening broader social and collective interests. The widespread popular political commitment to privatization adds a note of genuine urgency to this work.

Notes

1. The issue of local funding is also confused by the issue of state support. There is evidence that many states increase local aid to help offset federal cutbacks. More recent evidence suggests, however, that states have also begun to cut back direct local aid. For an excellent review of the literature describing federal cutbacks in the 1980s, see MacManus (1992).

2. Sometimes the link between "market forces" and policy can be very indirect. For example, the virtue of having employment training programs designed by Private Industry Councils is that it taps the knowledge of industry leaders about the labor needs of the local market. There are no direct market forces operating on the councils.

3. The issue of time is of particular importance. Critics argue that private vendors will sometimes forgo immediate profits to obtain market control. These vendors may be able to obtain "excess profits" from public authorities at a later time, because there are no alternative service providers.

4. The question of public employees is quite important. Many public choice theorists argue that the large growth of government over the past three decades is at least in part the result of the power of public bureaucracy to create political pressure to assure its continued growth (Buchanan, 1984).

5. See, for example, Stone's (1989) work on Atlanta and Elkin's (1987) work on Dallas. A number of shorter case studies are presented in Stone and Sanders (1987) and Judd and Parkinson (1990).

6. For description largely restricted to issues of economic development, see Fainstein and Fainstein (1986), Cummings (1988), Stone and Sanders (1987), Squires (1989), and Judd and Parkinson (1990). For broader treatment of local decision making, see Elkin (1987) on Dallas and Stone (1989) on Atlanta.

7. For an overview of the development of the notion of local political culture, see Clark and Ferguson (1983), Clark and Jeanrenaud (1989), and Clark and Inglehart (1990).

8. A significant body of survey data suggests that something like new fiscal populism is emerging in a number of advanced industrial societies (see Inglehart, 1987).

9. Throughout the 1970s, the number of city employees dropped. Overall spending has continued to grow, but its rate of increase has slowed significantly.

10. It is interesting to note that although the implementation of economic development policy through the use of such organizations is not uncommon, the organizations are not well discussed in the privatization literature.

11. The Greater Baltimore Committee is clearly the more powerful of these two groups. An explicitly elitist organization composed of the chief executive officers of the city's 100 largest businesses, it was established as an association of business interests dedicated to the economic revitalization of the city. The Committee for Downtown was an association formed by central-city merchants.

12. The limits of this influence are the subject of some debate. Levine (1987) argues that the city has essentially delegated its entire redevelopment program to the private sector. Berkowitz (1987) strongly rejects this view and claims that the city has maintained an assertive leadership role. Whatever position one takes, however, it is clear that the corporation's influence on redevelopment planning and capital expenditures is far from negligible.

13. One could cite a variety of additional examples of quasi-private corporations active in Baltimore's redevelopment program. Estimates vary, but most observers recognize 30 to 40 such organizations. Obviously, some of these are quite active and spend a good deal of money; other are much less important. For some of the organizations, budget and even function data are difficult to obtain. Whereas Baltimore Aquarium Incorporated oversees an operating budget of $650,000 and supervises a multimillion-dollar capital budget, others are paper corporations created to facilitate fund transfers. A complete description of the overall ecology of Baltimore's shadow government is even more complex than a listing of institutions would suggest. Numerous projects not directly administered by such agencies are often closely tied to the redevelopment network.

14. Given the changing demographics of the city, such findings should not be surprising (see Levine, 1987).

15. The debate surrounding the IBM building was of particular importance in that it called into question a number of long-held assumptions about the physical design of the Inner Harbor renewal. Specifically, it forced a reexamination of the principle that development near the harbor should be of limited height to assure visual access.

16. Those who doubt the utility of the concept of social interest can substitute "bureaucratically defined interest."

17. This finding should not be surprising. There is a good deal of evidence of such displacement in past public-private efforts. Of particular interest is the history of the urban renewal program. For a review, see Mollenkopf (1983).

18. This might seem to be an obvious point. Nevertheless, both practitioners and scholars have endorsed a wide range of privatization strategies on the basis of evidence generated by a very limited set of privatization schemes. Often only contracting is considered.

References

Arnold, J. (1982). The politics of the Baltimore renaissance. In L. H. Nast, L. N. Krause, & R. C. Monk (Eds.), *Baltimore: A living renaissance.* Baltimore: Historic Baltimore Society.

Benjamin, R., & Elkin, S. L. (1985). *The democratic state.* Lawrence: University Press of Kansas.

Berkowitz, B. L. (1984). Economic development really works: Baltimore, Maryland. In R. Bingham & J. Blair (Eds.), *Urban economic development.* Beverly Hills, CA: Sage.

Berkowitz, B. L. (1987). Rejoinder to "Downtown Redevelopment": A critical appraisal of the Baltimore renaissance. *Journal of Urban Affairs, 9,* 125-132.

Bonnell, B. (1982). Charles Center-Inner Harbor Management Inc. In L. H. Nast, L. N. Krause, & R. C. Monk (Eds.), *Baltimore: A living renaissance.* Baltimore: Historic Baltimore Society.

Breckenfeld, G. (1977, March). It's up to the cities to save themselves. *Fortune,* pp. 196-206.

Buchanan, J. M. (1984). Politics without romance: A sketch of positive public choice theory and its normative implications. In J. M. Buchanan & R. D. Tollison (Eds.), *The theory of public choice, II.* Ann Arbor: University of Michigan Press.

Clark, T. N., & Ferguson, L. C. (1983). *City money: Political processes, fiscal strain, and retrenchment.* New York: Columbia University Press.

Clark, T. N., & Inglehart, R. (1990, July). *The new political culture: Changing dynamics of support for the welfare state and other policies in post-industrial societies.* Paper presented at the biennial meeting of the International Sociological Association, Madrid.

Clark, T. N., & Jeanrenaud, C. (1989, April). *Why are Swiss leaders invisible?* Paper presented at the Conference on New Leaders, New Parties and New Groups in Local Politics, Fiscal Austerity and Urban Innovation Project, Paris.

Cummings, S. (1988). *Business elites and urban development.* Albany: State University of New York Press.

Donahue, J. D. (1989). *The privatization decision: Public ends, private means.* New York: Basic Books.

Eisinger, P. K. (1988). *The rise of the entrepreneurial state.* Madison: University of Wisconsin Press.
Elkin, S. L. (1987). *City and regime in the American republic.* Chicago: University of Chicago Press.
Fainstein, S., & Fainstein, N. (Eds.). (1986). *Restructuring the city: The political economy of urban redevelopment.* New York: Longman.
Fesler, J. W., & Kettl, D. (1991). *The administrative process.* Chatham, NJ: Chatham House.
Hero, R. (1987). The urban service delivery literature: Some questions and considerations. *Polity, 18,* 659-677.
Hula, R. C. (1990). The two Baltimores. In D. Judd & M. Parkinson (Eds.), *Leadership and urban regeneration: Cities in North America and Europe.* Newbury Park, CA: Sage.
Inglehart, R. (1987). Value change in industrial society. *American Political Science Review, 18,* 1289-1303.
Judd, D., & Parkinson, M. (Eds.). (1990). *Leadership and urban regeneration: Cities in North America and Europe.* Newbury Park, CA: Sage.
Levine, M. V. (1987). Response to Berkowitz, "Economic Development in Baltimore": Some additional perspectives. *Journal of Urban Affairs, 9,* 133-138.
Lipsky, M. (1980). *Street level bureaucracy.* New York: Russell Sage.
Lyall, K. (1980). A bicycle built-for-two: Public private partnership in Baltimore. In R. S. Fosler & R. A. Berger (Eds.), *Public-private partnership in American cities.* Lexington, MA: D. C. Heath.
MacManus, S. A. (1992, April). *Budget battles: The strategies of local government budget officers during recessionary periods.* Paper presented at the annual meeting of the Urban Affairs Association, Cleveland.
Mansbridge, J. J. (1990). *Beyond self-interest.* Chicago: University of Chicago Press.
Mollenkopf, J. (1983). *The contested city.* Princeton, NJ: Princeton University Press.
Osborne, D., & Gaebler, T. (1992). *Reinventing government.* Reading, MA: Addison-Wesley.
Pammer, W. J. (1992). The future of municipal finance in an era of fiscal austerity and economic globalization. In International City Managers Association (Ed.), *The municipal yearbook.* Washington, DC: International City Managers Association.
Peterson, P. E. (1981). *City limits.* Chicago: University of Chicago Press.
Pressman, J., & Wildavsky, A. (1984). *Implementation.* Berkeley: University of California Press.
Savas, E. S. (1987). *Privatization: The key to better government.* Chatham, NJ: Chatham House.
Smith, C. F. (1980, April 16). City rescue of Coldspring apparently broke U.S. regulations. *Baltimore Sun,* p. A-1.
Squires, G. D. (Ed.). (1989). *Unequal partners: The political economy of urban redevelopment.* New Brunswick, NJ: Rutgers University Press.
Stoker, R. P. (1987). Baltimore: The self-evaluating city. In C. N. Stone & H. T. Sanders (Eds.), *The politics of urban development.* Lawrence: University Press of Kansas.
Stone, C. N. (1985). Efficiency versus social learning: A reconsideration of the implementation process. *Policy Studies Review, 4,* 484-490.

Stone, C. N. (1987). Summing up: Urban regimes, development policy, and political arrangements. In C. N. Stone & H. T. Sanders (Eds.), *The politics of urban development* (pp. 269-290). Lawrence: University Press of Kansas.

Stone, C. N. (1989). *Regime politics: Governing Atlanta, 1946-1988.* Lawrence: University Press of Kansas.

Stone, C. N., & Sanders, H. T. (Eds.). (1987). *The politics of urban development.* Lawrence: University Press of Kansas.

Sullivan, H. J. (1987, November/December). Privatization of public services: A threat to constitutional rights. *Public Administration Review, 48,* 461-467.

Szanton, P. L. (1986). *Baltimore 2000: A choice of futures.* Baltimore: Morris Goldseeker Foundation.

Whelan, R., & Young, A. H. (1989, September). *The New Orleans urban regime during the Barthelemy years.* Paper presented at the annual meeting of the American Political Science Association.

3

Local Authorities and Economic Development in Britain

JAMES A. CHANDLER

At first glance it might be expected that British local authorities should be well established and effective agencies for promoting the economic development of their communities. In England and Wales the nonmetropolitan districts are the lowest tier with effective powers and resources; they administer populations ranging from 21,500 to 396,000 inhabitants. The upper-tier counties in England range in population size from 21,000 to more than a million and a half inhabitants. The large conurbations are divided into single-tier metropolitan districts with minimum populations of 158,000, whereas London is divided into 33 borough councils. Scotland is divided into nine large regions and 53 smaller districts, but even here the smallest council has a population of 10,000.

Local governments in Britain are, therefore, sufficiently large to be able to coordinate planning and development of their industrial and commercial bases and, potentially, have the resources to make an impact on their local economies. Unlike in the United States, most local governments also envelop the whole of an urban conurbation and can, therefore, develop plans on an economically and geographically viable basis. London is, however, a notable exception, being the one capital city in Europe not governed by a single strategic authority. In reality, however, British local authorities probably have less impact on their local economies today than they had in the late nineteenth century and generally trail behind the experience of city governments in the United States in stimulating economic growth.

The problems faced by British local authorities in realizing their potential for stimulating their local economies stem to some degree from the horizontal pattern of intergovernment relationships within the state. The weakness of horizontal linkages between local governments in Britain has not greatly facilitated the development of economic interventionism within local governments. As substantial, well-resourced organizations, local authorities in Britain developed an attitude that they were self-sustaining organizations that had little need to cooperate with either their fellow local governments or other external private or public sector agencies to achieve their goals. The typical local authority, until recently, expected to provide services, from initiating policy to implementation, solely in-house. Local authority associations developed into no more than politically divided interest groups to pressure central government for greater resources. There is, therefore, little tradition or capacity for British local authorities to develop economic initiatives through cooperation among authorities or with private sector organizations.

The greatest limitation on economic intervention by local governments lies, however, in the vertical relationship between central and local government. Central-local relations in Britain have been described as a partnership in which central and local governments have separate but complementary roles within the state and, in contrast, as an agency model in which local authorities are wholly subservient to the center. The partnership model is much the more difficult to sustain, given the frequent reordering of the structures and functions of the local government system in Britain this century. The agency model is, however, not wholly appropriate, because local authorities have considerable discretion within the legislative framework and funding limits established by Parliament.

A more apposite model of the relationship is of stewardship, in which central government allows local authorities considerable power and authority to undertake tasks that it considers to be valuable for the state but that are either too detailed or too centered on local factors to be undertaken by the central administration. The powers given to localities can, therefore, be substantial in specific areas and may be used, on occasion, to oppose central government. However, localities' capacity for recalcitrance can be used only to a limited extent. As demonstrated by the failure of radical-left local authority rebellions against the Thatcher government, the state has sufficient means and authority to rein in local governments that act against central interest. I will therefore show in this chapter that although local governments initiated many innovative methods of aiding local economies, their capacity to sustain these initiatives has always been subject to review by central government.

The Development of Local Economic Intervention

Serious enthusiasm for a local authority role in economic development reemerged in the late 1970s and has now become established as an accepted task of local government through the Local Government and Housing Act of 1989. Economic intervention was a rediscovered rather than a new role for local authorities. During the present century, local authorities have been shaped by central government into service providers whose primary role lies in supplementing the welfare state rather than in economic production. This was not the case in the late nineteenth century, when the governments of the large cities were controlled by local industrialists who were concerned to ensure municipal provision of inexpensive gas, electricity, water, and transport for their communities in order to give their businesses a competitive edge over rival firms in neighboring towns (Fraser, 1979; see also Jacobs, Chapter 4, this volume). They secured powers to fulfill this role through private acts of Parliament, because general legislation relating to local authorities did not give them any substantive legal rights to enable local economic sponsorship. As large firms expanded to operate on a wider, national rather than city, basis, business began to direct its attention toward central rather than local government. Direct economic intervention through municipal ownership of services such as gas, electricity, and transport stagnated and then was rolled back as much of this enterprise was transferred to state nationalized control and, in the 1980s, into private ownership (Chandler, 1988).

During the 1930s, the economic depression did, however, motivate a number of authorities to seek industrial development by establishing factory sites and petitioning central government for subsidies to enable them to attract incoming firms. Resort towns such as Blackpool had also established effective publicity offices and provided parks, piers, and promenades to capture the tourist trade (Fogarty, 1947). These initiatives were brought to an end with the outbreak of World War II and, in the immediate postwar years, central planning controls and, later, full employment further dulled the edge of economic initiative. By the mid-1960s some local authorities in relatively depressed areas such as Cleveland or Lancashire had, however, established small offices for attracting new industry based largely on advertising and development of factory estates (Chandler & Lawless, 1985). More active services for attracting relocating businesses were developed in the new towns, such as Milton Keynes, that were created outside local authority control in order to alleviate inner-city overcrowding.

Reawakening of Concern for Local Development

Decline in economic growth in the 1970s prompted several local authorities to expand their efforts to stimulate local industrial development. The county of Cleveland and the city of Bradford, for example, created economic development units to pursue more aggressive advertising techniques pioneered by the new towns. Local authorities also began to realize their importance as employers and purchasers in their own right. Several local authorities developed schemes to give preference to local firms in purchasing goods and services or to use their direct works departments as a means for increasing employment and training within the local building industry. More extensive cooperation with business developed in Lancashire and Yorkshire, where several communities based on wool or cotton trades formed joint committees with local textile employers to pressure the government for better trading conditions.

Particular impetus toward local economic intervention was, largely unwittingly, provided by the Conservative government of Edward Heath through the creation of the metropolitan county councils and Scottish regions. These authorities enthusiastically embraced their potential for economic intervention, partly to demonstrate their value as new public entities. In the late 1970s, Manchester and Merseyside County Councils under Conservative control began to realize their development potential by setting up agencies to fund local businesses; these became the nuclei for more extensive "enterprise boards" within the metropolitan counties when, after 1981, they returned to Labour party control (Coulson, 1990).

Private Sector-Sponsored Local Development

Attempts to rescue ailing local economies were not motivated solely by local authorities. In 1978 the major employer in the glassmaking town of St. Helens, Pilkingtons, developed, with the assistance of the local authority and the government's Manpower Services Commission, the first of many local enterprise agencies. This organization, the St. Helens Trust, inspired several other companies to sugar the pill of redundancy by developing local employment-creation schemes. Nationalized industries, including British Steel, established local development companies to help workers made redundant by closures of plants. Private-sector funding for local business initiatives crystallized in the formation of Business in the Community in 1981, a body funded by a number of employers to establish, along the lines of the St. Helens Trust, a new small business enterprise, aided both by funds and by commercial advice. The organization works today on many initiatives that are in partnership with local authorities (Moore & Richardson, 1989).

Radical Alternatives

In the early 1980s the rightward turn of the Thatcher government and the largely consequent increase in unemployment stimulated a move to the left among many Labour-controlled local authorities. While more politically centrist cities such as Bradford developed strategies to charm industry into their communities, socialist authorities began to develop policies aimed at demonstrating the incompetence of capitalism and its responsibility for the increase in poverty and unemployment. Leaders in the local socialist movement were the city of Sheffield, where the government refused further subsidies to declining heavy industries, and, following the Labour party's local electoral successes in 1981, the Greater London Council, where, despite affluence in Mayfair and Chelsea, inner-city transition zones housed the most deprived communities in the country.

The radical left-wing authorities of the early 1980s promoted policies that have been generically termed "local socialism" (Gyford, 1985), although many of these councils pursued rather different strategies and often emphasized one policy area as distinct from another. In general, however, there was consensus among them that the increase in unemployment was generated by the failure of capitalists under the management of a New Right laissez-faire government. The left maintained that local economies were being destroyed by market forces outside local control and were throwing on the scrap heap of redundancy great potential within a skilled, but underutilized, workforce. These authorities maintained that the unemployed could be coordinated to create wealth within their own communities. Optimistically, local leaders believed that they could help secure for the skilled unemployed the conditions and finances to ensure that they rebuilt the local economy and thereby demonstrate that local working-class initiative has more to offer than do absentee capitalists. The Greater London Council, typically for these authorities, sought not just to alleviate unemployment but also to transform capitalism. The council's aims included a desire to "increase the element of democratic control over industrial decisions [and] encourage new forms of social ownership" (Greater London Council, 1981).

Radical local authorities developed schemes for allocating funds to stimulate local businesses through a variety of strategies. Initially, the most important source of funding came from the use of Clause 137 of the 1972 Local Government Act, which allowed councils to raise a limited rate to undertake tasks for the good of the community as a whole that were not permissible under other legislation. In the early 1980s, the Greater London Council was able to generate some £8 million a year from such a source; they used these funds to form the Greater London Enterprise Board to distribute grants to local firms, with reference not only to commercial viability but also to the form of ownership of the business and its support for trade unions and workers' rights. Preference was given to cooperative ventures on a small or medium scale and also to companies employing ethnic or women workers. The business also had to recognize the rights of employees and manufacture "socially useful" products. Other local authorities followed the Greater London venture, including the West Midlands County Council and the county of Lancashire.

Following a somewhat different strategy, the city of Sheffield decided to fund new socialist business ventures without the creation of an enterprise board. It created the first major department within a local authority to concentrate solely on local economic initiatives to channel funding activities. Sheffield's employment department also initiated aggressive campaigns to persuade the government to modify its industrial policies. The city issued a report on alternative strategies for development of the steel industry. Similar advocacy and research were pursued by other left-wing councils. The Greater London Council, for example, surveyed important business sectors in London and also campaigned for an alternative redevelopment strategy for London's dockland.

The left-wing authorities also took further the strategies that enhanced the local economy through the purchasing power and personnel policies of the local authority. Liverpool City Council sought to boost employment and solve local housing problems by building massive new housing estates. South Yorkshire Metropolitan County Council heavily subsidized bus fares, partly on the grounds that it permitted better mobility for workers. Radical authorities also cast around for legal loopholes that would permit them to extend their direct involvement in the local economy through, for example, manufacturing their own building products or selling garden produce. They also developed contract compliance schemes to demand that any organization supplying goods to the local authority should adhere to codes of employment requiring payment of reasonable wages to workers, acceptance of trade unions, and production of "socially useful," by which was meant nonmilitary, products. In addition, they turned their attention to their own personnel practices and emphasized the need to employ more women and ethnic minorities on their staffs.

Whereas conventional local employment initiatives had private sector equivalents in organizations such as Business in the Community, there was a modest move to parallel radical local authority initiatives through the private funding of cooperative economic ventures. The cooperative movement in Britain fostered by retail shops and the Bank and Insurance Company provided means to fund limited numbers of workers who wanted joint ownership of their own start-up businesses. Great interest was shown by some local authorities in the cooperative industries established at Mondragon in northern Spain. In many of the radical cities efforts were made to develop close liaisons between these cooperative schemes and the economic development plans of the local authorities.

The revival of interest in economic development also led the left-wing councils to adopt some strategies already being formed by other more moderate authorities. Efforts were made, for example, to form sheltered workshops in which individuals with commercial ideas could seek to develop their initiatives in low-rent accommodations, with advisers on funding, accountancy, and business strategy on hand to help guide the infant businesses. New training schemes to aid redundant employees were also an important function of the employment strategies of these authorities.

Several of the ventures developed by the left were generally successful. The enterprise boards formed by the Greater London Council and the metropolitan counties were viable in providing an alternative source of funding for small and medium businesses, and attention to employment creation stimulated the development of factory premises, advice for new businesses, and training for unskilled or redundant workers. In Sheffield the resources that could be collected by the city were, however, insufficient to have any impact on the decline of employment in the localized steel industry, and the government was not to be turned by the pressure exerted from a provincial city council. Once legislation began to have an impact, the more lasting achievements of the radical councils were in more conventional areas of business promotion that did not challenge the basis of capitalist ownership.

Central Government Reaction

During its first years, the Thatcher government seems to have been largely nonplussed by the range of local employment initiatives being undertaken by local government and private business. The government showed little or no recognition of the schemes in the early 1980s. Its policies rested on the belief that free market forces should be allowed to kill off outmoded traditional industries and that these would be replaced by modern competitive firms if controls over new enterprise were removed. These values did not provide for public subsidies to local businesses, contract compliance, or preferential local purchasing.

The principal focus for regeneration of localities by the government was, therefore, on schemes to ensure deregulation of business in selected areas targeted for growth. An important element of this policy was the creation of urban development corporations (UDCs), which took over planning and development powers in selected areas of industrial decline from the local authority. The UDCs were provided generous funding to secure development through building new premises and infrastructure and providing rent and rates subsidies and some lessening of planning and other government regulations. They supplanted local authorities as the planning agencies for the designated areas. In 1981 UDCs were formed to regenerate the London and Liverpool docklands. The former scheme was initially spectacularly successful, but it could scarcely fail, given that it was established as an overspill area for the crowded city of London. The Liverpool project and later UDCs have had more limited success and have flourished as much by attracting commercial as by attracting industrial developments. In addition to the UDCs, smaller areas were made into enterprise zones, where incoming businesses could enjoy rates and rent subsidies and freedom from many controls on health and safety and training.

The UDC and enterprise zone schemes demonstrated the Thatcher government's hostility to local government initiatives. Rather than use local authorities as a means of generating local economies and building on their local knowledge and growing experience of this task, the government bypassed their efforts. In some areas, such as Sheffield and Middlesborough, UDCs were formed to supplant existing development schemes established by local authorities and business partners. In all UDCs the local authority was largely left out in the cold. Their management boards include token local authority members, but they do not form the majority of members and cannot determine the UDCs' activities. Most UDCs have, nevertheless, cooperated with local councils, but they are far from being under local authority control.

Certain of the more radical schemes developed by local authorities were wholly beyond the pale for the New Right-influenced government and were effectively terminated by central government. The highly undemocratic Local Government Act of 1986 prevented local authorities from engaging in political critiques of central government and therefore curtailed public efforts to develop local opposition to the government's industrial strategy. Contract compliance was also brought to an end for all but issues of ethnic discrimination. Legislation prevented local authorities from subsidizing their transport or housing budgets from the rates. More generally damaging were curbs on local authority spending and capital borrowing restrictions, especially on housing, that prevented local socialist councils from generating the funds required to stimulate their local economies. There are also limits on the extent to which local authorities can greatly affect skills within the labor market; during the Thatcher years much of the local potential for developments in this sector was transferred to the centrally directed Training Commission, and is now held by the Training and Enterprise Councils.

One of the first measures to restrict local authority activity in production was the 1980 Local Government Planning and Land Act, which, among its many strictures, ruled that local authority direct works construction departments had to make a profit and submit large tenders to competition with the private sector. It therefore became impossible for local authorities to stimulate the local economy by injecting capital into construction development implemented directly by the authority or a favored local company. This demand was further developed in 1988, when many more local authority services became subject to compulsory competitive tendering. Local authorities are consequently now forced to operate many services as if they were private companies without the capacity to generate a trading loss for a few years or to expand their business through national or international trade. They are, therefore, forced to accept the demands for profitability through competing in the open market but without any powers to use their entrepreneurial skills to expand into markets other than those directly tied to the needs of the local authority. Thus local government in Britain has little scope for developing the local economy through expansion of its own productive or service capacity.

A further blow to local authority efforts to create employment was the abolition in 1985 of the Greater London Council and the metropolitan counties. This measure was to a large extent politically motivated, as these authorities had become major critics of Thatcherism. Their demise left London without any strategic planning and development authority and cut the organizational base from some of the largest-scale local authority development schemes. The London boroughs and metropolitan district authorities took over many of the powers of the lamented upper-tier authorities and were able to keep some of these initiatives on a joint basis. The enterprise boards, in particular, were salvaged, although by no means all the lower-tier authorities participated in their retention; most of the boards consequently faced reductions in revenue from their sponsoring authorities.

While bypassing local authority-led employment initiatives, the government held back from curtailing all attempts by local authorities to create jobs and was generally careful not to deny that local authorities had legitimate concerns for their local economies. Suggestions made in 1982 in a consultative Green Paper that Section 137 money could not be used for local employment initiatives were not followed up in practice. Such action would present a publicity gift for aggrieved local authorities seeking to deal with unemployment, which was widely seen as one of the most unpopular consequences of Thatcherism.

By the late 1980s, central government had gained from previous experience in dealing with the local authorities and arrived at means whereby, rather than directly curtailing local government employment initiatives, they channeled them into directions that suited their free enterprise values. The 1989 Local Government and Housing Act, for example, recognized local authority powers to aid the economy, but the government also ensured that most practical schemes for major regeneration had to be developed with local authorities as junior partners with private developers. Any arm's-length body created by local authorities, such as enterprise boards, that contain a majority of local authority nominees must be regarded as part of the local authority and thus subject to funding and trading restraints placed on local government. This has forced many local authorities to either close or privatize their enterprise boards by surrendering their control to other nonpublic interests.

Local Authority Initiatives in a Cold Climate

Although the ambitious local attempts to restructure capitalism have run aground, there has been a steady growth in economic interventionism of more moderate design. In Sellgren's (1986) postal survey of British local authorities, 118 responded with examples of economic initiatives. These schemes were predominantly centered on provision of advice and consultancy, land, and premises that were established in the 1960s and 1970s. Many local authorities in more economically depressed areas continue to attempt to attract mobile industry through advertising campaigns and promises of infrastructure improvements. This practice is still permitted by central government in preference to other economic improvement schemes, even though it is widely criticized as a wasteful practice of sterile competition among needy areas (Urry, 1990). Resort towns also pursue their attempts to attract vacationers and have been joined by some unlikely industrial towns eager to improve their images as centers for attracting clean modern industries. Some local authorities also continue to provide modest schemes to facilitate training for their local workforces through locally controlled colleges and a few sponsored training workshops. The Sheffield City Council, for example, provides facilities specifically for training women in electronics.

Sellgren's survey found that the majority of these activities were sponsored and funded principally by the local authorities rather than through partnership schemes, which are generally undertaken by a smaller number of larger authorities. This tendency is a reflection of the relatively small-scale nature and established values of many current local authority employment initiatives. Many of the more effective interventionist policies have been thwarted by legislation since 1979; for many smaller local authorities, the economic development policies of the 1990s have been molded by central government controls to reflect the concerns of the 1960s rather than the expansionist enthusiasm of the early 1980s.

The Development of Public-Private Partnerships

Although many radical ideas for local economic development have been downgraded into small-scale sterile competition for mobile industries, larger local authorities still have some stake in grander schemes of economic regeneration. These strategies, usually involving a package of financial grants, premises, and professional support, are increasingly public-private partnerships. Smallbone (1990) shows that in London, 18 out of 24 local enterprise agencies had local authority involvement, but in 11 of these the local authority was more a sponsoring body than a fully active partner in the project. Birmingham has forged a major scheme for renovating a large area of dereliction in the city by developing a sponsoring agency in partnership with several major construction companies. Although the Birmingham City Council was instrumental in setting up the scheme, it cannot be said to be the prime policy maker within the structure. Many of these larger ventures are, however, established initially through private sector initiatives involving groups such as Business in Industry, and some may dispense with any local government involvement, such as the London Enterprise Agency, a private sector alter ego of the London Enterprise Board. In general, however, whether established by private or public initiative, the tendency is for such schemes to have local authority input that facilitates the necessary planning consents, infrastructure improvement, and often land, whereas the private sector lends finance and commercial advice and expertise.

During the latter half of the 1980s, the ambitions of the left-wing policies of the "local socialist" authorities were also toned down and molded by central government into strategies based on cooperation, in junior partnership, rather than confrontation, with the private sector. The Sheffield City Council, for example, stopped trading insults with local business through the columns of the press and, in 1985, combined with the local chamber of commerce to create a jointly sponsored scheme to redevelop an area of the city left derelict following the collapse of steel plants. This joint project was subsequently taken over by a government UDC but had generated a successful working partnership between the Labour Authority and local businesses. The city council then successfully bid to stage the World Student Games as a prestige project to attract interest in the city and make it a center for sport. This project has involved extensive development of sporting facilities in partnership with private sector organizations. The leader of the council accepts that such a liaison may not lead to the developments favored by radical Labour party politicians and realizes that without such cooperation other ideas to regenerate the city would be unproductive. He argues that he could not pursue "the only alternative," which would have been "to sit on our hands and do nothing."

Attempts to create businesses that were based primarily on alternative forms of ownership such as cooperatives are also much less in evidence. It is clear that commercial success is possible only in firms that have not only skilled workforces but also the business knowledge and contacts to be able to fund and market their products. The local authorities that criticized the business sector for the slump of the early 1980s have, in large measure, come to realize that without the help of the commercial sector little economic progress is possible under prevailing economic conditions. Indeed, in many cases there has been some realization among erstwhile socialist councillors that some private sector firms may be more open to persuasion concerning the social costs of development than some public bodies, in particular, development corporations and central government. Local authorities have found allies in some private development companies in pushing the London Docklands Urban Development Corporation into guaranteeing jobs and housing for local people in the development.

Impact of Local Economic Development Initiatives

Many local authority employment departments have made estimates of the numbers of jobs created by their enterprises. Sheffield, for example, claimed in 1978 that it had created 5,400 new jobs in the city (Chesworth, 1989). The St. Helens Trust estimated that between 1977 and 1983 it created some 5,000 jobs (Moore & Richardson, 1989). These are, of course, highly tendentious estimates, and it is clearly difficult to ascertain whether many new jobs in a city would have arrived through the operation of an unfettered market rather than as a consequence of local or national economic incentives. Local authorities do not claim that their efforts are the solution to unemployment in Britain, but they argue for their value in economic development on the grounds that they are much more cost-effective in creating jobs than is central government. It is estimated by Business in the Community that the cost of creating jobs through local enterprise agencies is a fraction of the estimated cost per job for central government regional aid schemes (Moore & Richardson, 1989). Similar statements are made by local authorities (Chandler & Lawless, 1985, p. 246). Although these claims cannot be assessed accurately, they probably have considerable substance and have not been challenged seriously by central government.

It is, however, clear that local governments do not have, under their present powers, the resources in terms of finance or planning powers to make a huge impact on an area in decline unless they team up with major private sector interests or the government. The transformation of the London docklands would, financially and legally, have been beyond the powers of the London boroughs and, without government subsidy, perhaps even the now defunct Greater London Council. Acting on its existing resources a large local authority can, at best, hope that its effort to promote a community economy is the marginal factor that determines a major manufacturer to locate in the area. The resources available to fund new businesses, create premises for small- to medium-sized firms, or establish sheltered workshops are too small to create but a few hundred jobs. Nevertheless, these jobs are valued by those who receive them and, according to the local authorities, are achieved at lower cost than employment schemes sponsored by central government.

Even though local authorities may lack the resources to have a major independent impact on local economic development, they have considerable potential for serving as a catalyst to promote the urban growth coalitions necessary for economic expansion. Many local developments have been initiated by local authorities, and most private schemes for major regeneration of derelict areas have found it expedient to involve local government, at least as a junior partner, early in the project. The larger strategic authorities in Britain are well placed to fulfill this task and, hence, county and metropolitan district governments, in the absence of regional structures in England, have an important role to play in employment creation.

Local Government Initiative and the Political Environment

The efforts of local governments to generate economic resurgence in their areas clearly fall under what Peterson (1981) describes as developmental policies that attempt to improve income for a locality and benefit all groups in the community. In such circumstances there is little likelihood of serious dissent over the principles of pursuing such policies in local authorities with a declining economic base. Both local industrialists and professional and manual employees will see potentially positive benefits for themselves through successful pursuit of development policies.

Within the less affluent local economies there is, therefore, little internal dissent between political parties concerning the development of policies for economic regeneration when this is aimed at bringing new business into the area. There is, however, likely to be some dissent should interests consider that the means being used are unlikely to provide value for money. A highly costly strategy, such as Sheffield's development of sports facilities to stage the World Student Games within the city, draws as much criticism for its costs from within the ruling Labour party as from the local Conservative party.

Internal dissent is also likely if a local authority pursues a developmental policy that is anticapitalist and concerned not only to enhance employment, but also to change business ownership. Radical schemes pushed forward by optimistic socialist councils in the early 1980s have virtually disappeared, largely as a result of the impossibility of pursuing major socialist change in predominantly capitalist societies from the base of a beleaguered local authority. Socialist authorities faced the problem of raising capital for any cooperative businesses of significant size, because the firms themselves had little potential for attracting private capital. These authorities also suffered, to some extent, from a lack of knowledge concerning marketing and accounting and general management techniques. In radical Labour cities, redistributive strategies may have foundered even without the hostility of central government.

A further problem that has bedeviled employment creation is the lack of traditions of horizontal linkages among local authorities themselves and between local governments and the private sector. Most of the more innovative authorities in the early 1980s attempted to develop their schemes in isolation from one another and, more damagingly, without substantial enthusiasm for cooperating with industry. However, during the decade many local governments showed considerable competence in learning from early mistakes. Some cooperation between local governments is secured through the jointly funded Centre for Local Government Economic Strategy, which now advises on best practices within local government. The growth of partnership schemes demonstrates the capacity for local authorities to forge ahead along not well-traveled paths once their tendency to provide all services solely from their own resources was curtailed.

The Legacy of Thatcherism

The greatest potential threat to local economic initiatives, apart from the avowedly socialist schemes, comes from central government attitudes to local authorities in general and the New Right Conservative suspicion of collective enterprise in particular, rather than any local opposition to employment generation.

During the twentieth century, central governments in Britain have insidiously redirected the functions of local authorities from being largely devoted to developing the infrastructure and productive capacity of local economies toward being residual providers of social welfare (Dunleavy, 1984). This trend has been exacerbated by a prevalent view in Whitehall and Westminster that local authorities are the stewards of central government: entities given powers to undertake specific tasks required by the center but not encouraged to innovate into areas outside their assigned role (Chandler, 1991).

These trends have been fostered by both Labour and Conservative administrations. Between 1945 and 1951, for example, the Labour government removed local authority influence over gas and electricity production through nationalization of these industries, and in 1973 water became a regional rather than local concern. The creation of new towns by the Attlee government was pursued outside local authority influence. Thus efforts of local authorities to intervene in local economies have constituted a move that counters the trend toward greater centralization of local activities and the development of local authorities as service providers.

These tendencies were further exacerbated by the policies of the Thatcher government, which revolutionized British politics through its enthusiasm for rolling back the frontiers of the state. The Conservative governments also fully accepted the stewardship ethos and hence expected that local authorities would curtail their activities as part of the process of cutting back the public sector. The government did not, therefore, anticipate any expansion in local government activities, let alone into the area of economic development, which was an area best left, under New Right values, to the private sector.

Under these circumstances it is perhaps surprising that the Conservative party for much of the 1980s tolerated local authority economic development. Conservatives occasionally cast critical glances at the progress of development, and in a Green Paper of 1982 they suggested curtailing expenditure in this area. The government was, however, in rather a dilemma on the issue. Although it disapproved of the local initiatives on ideological grounds of laissez-faire and as a move by local government into productive activities outside central control, the government probably realized that legislation curtailing such activity would be electorally difficult to justify at a time of recession. Moreover, the widespread consensus between workers and industrialists favoring affordable local development policies would be difficult to curtail. A further factor deterring immediate government action was the volume of legislation molding the activities of local government that required action and parliamentary time. There were easier political targets to attack before launching the weight of government against more popular policies.

However, during the 1980s the government by various means effectively channeled local authority employment creation into an enabling rather than proactive role that generally ensures that local authorities can achieve only minor economic gains without private sector cooperation and partnership. The Thatcher government even suggested that it was supportive of local authority economic activity by explicitly recognizing this work as a legitimate activity for local government in the 1989 Local Government and Housing Act. This acknowledgment of reality was conceded only following a process of ensuring that local governments had relatively limited involvement in their local economies without the consent of other organizations. This has principally been secured, as described earlier, by restructuring larger active authorities, curtailing local expenditure, and preventing local authorities from avoiding fiscal controls through arm's-length enterprise companies. The consequence of these moves, along with general fiscal restraint, has been that in the 1990s most local authorities are assuming traditional roles of promoting economic development. They are given the opportunity to develop partnerships, usually as junior members, in economic developments; they have powers over infrastructure development, often subject to government capital allowances and planning consents; and they may publicize the attraction of their areas to industry.

Conclusion

Although nineteenth-century municipalities in Britain were much involved in local economic activities, the centralizing tendencies of British governments gradually made such activity a marginal local authority function. Economic decline from the 1970s and a brief flowering of radical municipal socialism in opposition to the Thatcher New Right generated a revival of local interest in aiding ailing local economies.

The impact of this renewed economic interventionism should not be measured simply in terms of jobs created. The 1980s revival of enthusiasm within local authorities to help shape their local economies was one of the few positive gains for local governments in a decade in which they were subjected to continual buffeting by a hostile centralist government. In the broadest sense, the development marked the revival of a new role for local authorities who do not perceive themselves as solely organizations fulfilling niches within the welfare state by providing services to the public. After many decades in which they have been shaped so as to be unrepresentative of community action, some local authorities are now aware that they may have a role to play in guiding the economic and social development of their communities and representing their communities' aspirations.

The role of local government as the center for growth coalitions has been recognized reluctantly by central government in its recent acknowledgment that local authorities can have concerns for the health of the local economy. This concession has been achieved through the revival in some cities of the spirit of municipal enterprise that created the magnificent civic pride and cathedral-like town halls of late Victorian industrial cities. Unfortunately, at the same time local authorities woke up to their potential role in improving their local economies, central government further turned the screws on their potential for realizing these newfound aims.

References

Chandler, J. A. (1988). *Public policy making for local government.* London: Croom Helm.

Chandler, J. A. (1991). *Local government today.* Manchester: Manchester University Press.

Chandler, J. A., & Lawless, P. (1985). *Local authorities and the creation of employment.* Farnborough: Gower.

Chesworth, M. (1989). *Sheffield city council's reaction to industrial change since 1975.* Unpublished bachelor's thesis, Sheffield City Polytechnic, Department of Public Sector Administration.

Coulson, A. W. (1990, May/June). Economic development: The metropolitan counties 1974-86. *Local Government Studies,* p. 103.

Dunleavy, P. (1984). The limits to local government. In M. Boddy & C. Fudge (Eds.), *Local socialism?* London: Macmillan.

Fogarty, M. P. (1947). *Plan your own industries: A study of local and regional development organizations.* Oxford: Basil Blackwell.

Fraser, D. (1979). *Power and authority in the Victorian city.* Oxford: Basil Blackwell.

Greater London Council. (1981). *Minority party report: A socialist policy for the Greater London Council.* London: Author.

Gyford, J. (1985). *The politics of local socialism.* London: George Allen & Unwin.
Moore, C., & Richardson, J. J. (1989). *Local partnership and the unemployment crisis in Britain.* London: Unwin Hyman.
Peterson, P. E. (1981). *City limits.* Chicago: University of Chicago Press.
Sellgren, J. (1986, November/December). Local economic development and local initiatives in the mid-1980s. *Local Government Studies,* pp. 51-68.
Smallbone, D. (1990, September/October). Enterprise agencies in London: A public-private sector partnership? *Local Government Studies,* pp. 17-32.
Urry, J. (1990). Conclusion: Places and policies. In M. Harloe, C. Pickvance, & J. Urry (Eds.), *Place, policy and politics* (pp. 187-204). London: Unwin Hyman.

4

Birmingham

Political Restructuring, Economic Change, and the Civic Gospel

BRIAN JACOBS

This chapter is about political restructuring and economic change in Britain's second largest city, Birmingham. The focus is on the impact of change upon the city's present ruling Labour administration. Socialist politics were adapted as municipal policy-making autonomy was eroded by economic restructuring and central government policies, forcing old-style "municipal socialism" to give way to a new "market socialism" (Gould, 1989).

From Civic Gospel to Market Socialism

Birmingham's political tradition of Labourism was characterized by both adherence to the development of the city's civic status and pride in its distinctive provincial cultural complexion. The pragmatic Labourism of the 1990s was not simply a reaction to the era of Thatcherism. It had been nurtured from the time of the birth of the national Labour party in 1906 by nonconformism and Fabian gradualism. It was founded in

AUTHOR'S NOTE: My thanks to Dr. James Radcliffe of Staffordshire University for items of data contained in this chapter.

the aspirations of the early British socialists, who had little enthusiasm for European-style Marxism (Cole & Postgate, 1966).

Birmingham has a special place in the history of local government in Britain. During the last century it was noted for its dynamic, achievement-oriented "civic gospel." In the Victorian age, civic pride was concerned as much with politics as with economics. The fruits of the civic gospel included a sweeping reorganization of the functions and finances of local government and the rapid development of a modern infrastructure of buildings and transportation (Briggs, 1963, p. 184). Birmingham, like Manchester, was an international city in terms of both its economic performance and its political and social magnetism. It was a nineteenth-century leader in technology, research, and development. Its surrounding "Black Country" was the "powerhouse" of the Midlands, known internationally as the "workshop of the world." It was the "city of a thousand trades," with a radical brand of politics and strong municipal drive.

The Victorian Vision

In many ways, Birmingham is still influenced by the Victorian vision. The civic gospel provided the city with a lasting optimism and confidence that has been noticeably absent in other cities in Britain. Birmingham's industrial and trading roots gave it a strongly working-class social base. In the late nineteenth century, manufacturing provided employment for 75% of the workforce, which was extensively unionized and politically oriented to the Labour party. Anti-Tory feeling "could easily be mobilized in Birmingham" (Briggs, 1963, p. 203). This, blended with a radical faith in city expansion, produced a strong civic leadership advocating a combination of market capitalism and municipal interventionism. The clearest expression of this came during Joseph Chamberlain's period as Liberal mayor in the 1870s. It was then, prior to the Labour party's arrival on the political scene, that there was a concerted implementation of the new civic gospel.

Chamberlain's vision was one of leadership to serve the people. His commitment was cast in stone by the erection of new civic buildings that were seen as rivals to those in Manchester, Leeds, and Glasgow. Chamberlain's era was one of exciting change that reorganized the spatial arrangement of large parts of the city and radically reformed the way in which local government intervened in the dynamic of change in Birmingham during those years. Chamberlain said that after he had served 12 months in office Birmingham would be "not knowing itself." Indeed, the civic gospel was implemented with a religious zeal that, in the early twentieth century, was reflected in Labour's commitment to "municipal socialism." Nineteenth-century municipal projects were implemented under the banners of progress, competition, and city corporatism.

Later economic decline created doubts about the effectiveness of both Labour and Conservative political leadership. Although the Victorian notion of "positive municipal action" survived, by the early 1980s Birmingham was "second city" more in name than in reality, its hinterland and central core having been ravaged by plant closures and unemployment and the thrust of economic growth having moved decisively to the southeast of England. The city's infrastructure went into decay, and its stable social base was undermined by racial tensions, unemployment, and economic deprivation. The confidence of the Chamberlain era gave way to a loss of political direction that persisted even following the transitory "boom" in the 1960s when Birmingham was buoyed up on a wave of speculative property development and expansion in the motor industry.

Embracing the Market

Municipal socialism in Birmingham had never been about *Das Kapital*. It was about the achievement of harmony between social classes rather than class conflict and social revolution (Cole & Postgate, 1966, p. 249). It was about being anti-Tory and against overcentralization, devotion to the industrial ethic of honest work for fair pay, and dedication to the evolutionary road to socialism.

Radicalism in Birmingham also had an "imperial reference" (Briggs, 1963, p. 239). Municipal socialism was connected with the future of the city as an important hub of empire trade and manufacture—an industrial center of international importance. Support for municipal prosperity legitimated the Labour party as an organic part of the provincial political culture. In the post–World War II era, Labour could thus join the Conservatives as a "natural party of government" (there were 68 Labour councillors, 37 Conservatives, 10 Liberal Democrats, and 2 others on the city council in early 1991).

From 1984, under the Labour group leadership of Sir Richard Knowles, the city avoided "purist" socialist ideology by following a business-oriented economic growth strategy (Birmingham City Council, 1986).

The new approach was to involve business with community, charitable, and other nonprofit organizations to develop the market and achieve socially responsible policies. Knowles referred to the days of past Victorian dynamism in the city to justify the council's proactive approach in drawing local organizations in to support Birmingham's vision of the new future. Ironically, it was a vision that appeared to complement Prime Minister Thatcher's desire to return to "Victorian values" as a way of generating enterprise in the cities. Some community leaders were therefore suspicious of the partnership line, especially when it was elaborated by Knowles, whom they expected to stand for "community" rather than corporate interests. Knowles (1989) justified the council's approach, but in terms of its past traditions and its old liberal radicalism, rather than through a specifically socialist appeal (see also Birmingham City Council, 1990).

Renaissance or Decline?

Under Knowles, the Labour council's high-profile, market-oriented, Economic Development Committee adopted a strategic vision that sold Birmingham to inward investors as Britain's "second city," full of flair and enterprise. However, the promotion of the city as a growth point presented problems (Longbottom, 1992). Despite important sectoral growth in the local economy, the city council presided over an economy that in many respects was in relative decline when compared with other U.K. and European cities. In political terms, the continued decline in key sectors during the 1990s recession raised doubts about the ability of the Labour administration to resist effectively the downward pressures of a recessionary market.

The Labour party endorsed a European-style urban renaissance. In their important book published in 1992, internationally renowned architect Richard Rogers and Labour's shadow minister for the arts, Mark Fisher, presented an impassioned plea for a future Labour government to use public intervention to achieve this. They argued that councils such as Birmingham, in the face of tough market conditions and a hostile centralizing government, could not possibly achieve the demonstrably successful and dynamic urban economic development that had been the hallmark of some European social democratic governed municipalities. Birmingham could not effect the kind of civic renaissance evident in cities such as Barcelona and Rotterdam unless public initiative was forthcoming (Rogers & Fisher, 1992).

Regional Decline

The problems facing Birmingham were sharpened by its sheer size and poor competitive position. With a population of 998,200 in 1988, Birmingham forms part of a larger and economically weakened conurbation that includes Wolverhampton, Dudley, and other towns in which Coventry and Birmingham, together with the rest of the conurbation, are often confusingly referred to as the "West Midlands County" (after a "top-tier" local government jurisdiction that was abolished in 1986). The "county" had a population of 2,625,000 in 1987, making it about the size of the Cleveland-Akron-Lorain consolidated metropolitan statistical area (the twelfth-largest U.S. conurbation) and a little smaller than Miami-Fort Lauderdale.

Within its even wider hinterland known as the West Midlands "standard region" (WMR), Birmingham retains its status as the "commercial capital." The city lies at the heart of England's motorway network and has an international "gateway" airport. It enjoys easy access to coastal ports and is a location favored by many national distribution companies. However, there is intense direct competition from London to the south and from Manchester to the north, with their well-established and expanding financial services and high-technology sectors. The competition among cities, even among those within the WMR, for inward investment indicates the difficult market conditions affecting the city.

Political Restructuring

Political attention was therefore concerned with attracting domestic and international companies to the city and relocations from other areas. Things seemed to go well as the recession in the early 1980s gave way to the Thatcher boom, but whether a new takeoff had been achieved was questionable. This was particularly so as the recession in the early 1990s again hit the West Midlands severely. There were conflicting signals in the 1990s from different sectors of the local economy that pointed to growth in service and high-technology industries and the continued decline and vulnerability of the manufacturing sector and a slowdown in property and construction. Service expansion threatened Labour's traditional base of support as the nonunionized labor force expanded, but it also provided new opportunities for labor to adapt to changed circumstances.

The old certainties of working-class labor politics were challenged as trade unionists lost their jobs and the number of persons in employment in the WMR fell below the 100 level index in 1988 (based on 1980). While the middle-class "yuppies" of the high-tech industries prospered, an underclass emerged in the inner city (Jacobs, 1992). In addition, the average unemployment rate fell briefly at the end of the 1980s, but so did per capita income between 1985 and 1987. Gross domestic product (GDP) in manufacturing industries in the WMR remained roughly constant between 1984 and 1988 (an important indicator of the region's generally slow growth), and average gross weekly earnings continued to decline sharply for both males and females. Housing starts remained flat, being well below the national average even during the 1980s property boom.

The GDP of the WMR in 1987 was estimated at £28.8 billion, or 8.4% of the U.K. total (excluding the continental shelf). The WMR's GDP per head in 1987 was 5,549, which was relatively low at 92% of the U.K. average. However, the four years up to 1989 showed some improvement, but this followed the long-term decline in the region's relative per capita GDP to 1983. In 1987, manufacturing was down to only a third of GDP in the WMR (although this was compared with a quarter of GDP nationally; it still represented the highest proportional share out of all the U.K. standard regions). Indeed, the postrecession period after 1981 saw this share of manufacturing stay at about the same level (Central Statistical Office, 1989, p. 19), despite the expansion of Birmingham's service sector.

Birmingham suffered from a decline in its population during the 1980s, persistent high unemployment, and acute social deprivation in housing, education, social, and health services. The Labour party thus developed a double-edged political response reflecting a need to appeal to those who had benefited from the Thatcher boom (skilled employed workers and the middle class) and who lived in the more prosperous suburban areas. It also represented a need to convince the underclass that their interests lay not in opposing middle-class prosperity but in backing inner-city enterprise and civic regeneration.

Autonomy and the Market

Economic change and political restructuring were thus intimately linked. Central government spending cutbacks and the pace of decline in the city and the region were issues at the forefront of the Labour party's political campaigning throughout the 1980s and early 1990s. Labour's claim was that government policies were undermining the true spirit of enterprise and this in turn harmed entrepreneurs and reduced the potential of the local business community (Knowles, 1989).

During the 1980s Birmingham was affected by a substantial reduction in the powers of local government and by Thatcher's municipal privatization (Pipe, 1992). Decision making was recast in the face of economic market changes affecting the relationships between the city's political decision makers and external corporate interests. The city thus lost some of its autonomy in terms of both its relations with the center and its ability to pursue its own interests independent of "external" economic decision makers, what King and Pierre (1990, p. 2) refer to as examples of losses affecting the two most critical elements that make up "local autonomy."

The city council intervened in the market by involvement with a number of subsidiary and associated companies, each of which played a strategic role in economic development within and around the city.[1] The council's involvement was influenced by the desire to maintain a direct local authority stake in urban renewal and economic development. However, council effectiveness was constrained both by the government's general suspicion of local authority companies and by its desire to reduce municipal interventions of this kind in favor of the corporate private sector.

Council policies in urban renewal and economic development had to be increasingly market led. This included the regeneration of the city's central business district (CBD) as a catalyst for development. The "internationalization" of the CBD involved the council's working with the European Commission, major development, property, and foreign service sector companies. There was also the hosting of important international sports gatherings and the holding of commercial and other events in the city (for example, at the International Convention Centre, the city's two universities [Game & Prior, 1991], and the National Exhibition Centre). The council also encouraged events that had high media profiles and that would attract international coverage. For example, in June 1991, the queen opened the International Convention Centre, which was simultaneously hosting the International Olympic Committee (IOC). International media coverage of the IOC was combined with reportage of the queen's visit to the city's impressive new Symphony Hall, home of the Birmingham Symphony Orchestra.

The New Locus of Power

Community group coalitions attempted to lobby the council and property developers to change proposed plans for key CBD sites. The "People's Plan" produced by the highly professional group Birmingham for People for the Bull Ring development in the city center persuaded the council and developers to acknowledge citizen group representation and the demands of market traders (Cowen, 1990; Holyoak, 1991). The plans for a "mega" shopping mall in the middle of the city on the site of an existing outdoor market created a tide of opposition that led the developers to submit a more "humane" design incorporating street-level open spaces together with an accessible new light rapid transit system.

Nevertheless, decision making on planning issues remained heavily influenced by the property development companies. The major companies that worked closely with the council were based outside Birmingham. They employed professional architects and designers to create projects to which the council responded through its planning committee; but the plans were often produced in London design offices or were localized variants of "off-the-shelf" schemes that could be "sold" to the public and then built quickly along "fast-track" lines using the latest American-imported know-how (Rabeneck, 1990). Such developments meant that the city council's planning committee increasingly operated as a facilitator for the private sector in conjunction with its regulative role in planning control (Wray, 1988). Although the planning committee could stipulate that developers should conform to council planning guidelines and could encourage property companies to change their plans, this was achieved largely within the parameters defined by economic and income-generating considerations.

Market-led development benefited the city as developers signaled their desire to purchase sites for retail and commercial development. During the 1980s, before the property slump, city-center sites were attractive because they were cheaper than locations in London and provided the prospect of fast returns. But the cost to the city was in terms of its lack of bargaining power when set against the big property development companies and international design teams.

There was good profit potential in the Birmingham office market despite the national downturn in 1990 (Stringer, 1990). The prime movers in the market were companies such as MEPC, with its Cornwall Court development; Speyhawk, with its large Snow Hill scheme; Henry Boot at Paradise Circus; Merlin Sherwater Laing, with proposals for a development near the International Convention Centre; Slough Estates and its out-of-town Kingsgate business center; and Ladbroke City and County Land, with the Perry Barr shopping center. These represented just a sample of the companies involved in Birmingham. They had national and international property interests and together were making a major impact on the spatial character and planning quality of the city. They were joined by international banks, service sector companies, and high-technology firms that had either expanded in the city or moved to CBD and out-of-town locations (Leppard, 1990; Stringer, 1990).

Apart from the combined impact of the financial investments of these companies in Birmingham, they exerted an influence on the nature and pace of urban development in a setting where the council was increasingly reactive in its role as planning authority. The sheer size and complexity of new developments required that the technical, economic, and market knowledge of the development companies be brought to the fore in putting together development packages (Harding, 1990). Speyhawk's proposals at Snow Hill, for example, were modeled on large-scale office schemes in Stockholm and Amsterdam. Speyhawk would create a "complete business environment" with a Phase 1 provision for 475,000 square feet of office space. Occupiers were promised access to a wide range of central services, presentation facilities, and television conference facilities. The scheme was centered on a galleria-style internal street running north to south and connecting retail units to a health club, restaurants, a concert area, and a computer services center. The Snow Hill galleria mall was echoed in other CBD schemes, such as the National Exhibition Centre, the Birmingham central library complex, and a variety of shopping centers.

Developments by the private sector were obviously welcome in terms of the benefits they brought to a city facing economic problems and unemployment. The political considerations were, however, profound, because they involved a discernible shift of influence and power away from the council toward the corporate private sector. The Victorian city fathers were concerned with establishing a strong municipal presence in conjunction with the private sector. However, in the 1980s the location of corporate decision making had been increasingly internationalized. The resources for investment deployed by companies were substantial and originated outside Birmingham and other provincial centers. City urban program financial resources looked pathetically small when compared with the sums raised in London, Frankfurt, and New York by corporate traders and development companies.

Technology allowed companies and banks with Birmingham branch offices, such as Sumitomo (the first Japanese bank to open in Birmingham in 1985), Tokai, Mitsui, Yokohama, Sanwa, and the Bank of Tokyo, to maintain and develop their communications networks on an international basis. The Japanese banks now located in the city could keep in contact with London, New York, and Tokyo in a way that was unimaginable in Joseph Chamberlain's day. The new technological infrastructure established by the banks and business enterprises was no longer bound by the confines of the city center or the vagaries of the public communications system. Nor were they dependent upon the trade-unionized labor force of the past that had been the bedrock upon which Birmingham's economic development had first been built.

In contrast, City 2000, representing 160 member firms, acted as a pressure group promoting Birmingham's indigenous commercial and business interests (Smith, 1991). This group also reflected the changes in the city's economic structure and the diverse nature of the local business community. It had a strong financial and professional membership that complemented the representation of manufacturing and commercial companies in the local chamber of commerce (with its 5,500 member companies).

Both City 2000 and the chamber could claim to be influential in relation to the interparty political decision-making process of the city council. They were interest groups that were respected and acknowledged by local politicians and officials and that acted as lobbying organizations, with members represented on various public and quasi-public bodies. The chamber was an active participant in the council's urban partnership programs and took part in managing and marketing the National Exhibition Centre and the International Convention Centre (Birmingham Post and Mail, 1991, p. 71).

Where the chamber's interests often diverged from those of the international corporate private sector (especially on small business issues), it tended to underscore the general externalization and globalization of influence over strategic decisions in the city. This, in turn, emphasized the need for the city to be clear about the economic priorities it had defined for itself and the city's indigenous commercial sector.

Community Empowerment

These initiatives did not automatically ensure the empowerment of distressed communities. Birmingham faced an acute inner-city crisis, with economically poor and politically weak communities juxtaposed against the background of the gleaming new retail and office developments in the city center. If the city was to recapture its international reputation, it had to commit substantial resources to improving both its inner-city residential areas and the quality of leisure and entertainment in the central business district, but was this to be at the expense of ensuring the involvement of local people in decisions affecting their lives?

Community groups illustrated their lack of influence by reference to their isolation from city-level decision making (Jacobs, 1986, 1988) and to their economic weakness exemplified by the depressing picture of the city's inner urban problems (see Table 4.1). The Birmingham Inner City Partnership (1989) highlighted the relative decline of distressed communities clustered around the central business district. These were Handsworth, Soho, Aston, Ladywood, Nechells, Washwood Heath, Sparkbrook, Small Heath, Sparkhill, and Acocks Green. All were noted for their high levels of crime, environmental neglect, racial tensions, and social distress.

In Handsworth, there were serious riots in 1981 and 1985 that led to conflicts between black and white rioters and the police. The 1985 disturbances were particularly violent. On September 8, 1985, black youths threw stones at police and used petrol bombs. Police were deployed with full riot gear as youths attacked shops and commercial premises and burned down a number of buildings. After the riots the devastation was extensive. Asian and black businesspeople complained about the neglect of the locality and the lack of resources available from public sources to rebuild (Jacobs, 1986, 1988). The riots generated

Table 4.1 Selected Districts Showing Indicators Using Regional Trends Data, 1989

District	*% Population Change, 1982-1987*	*Deaths per Thousand, 1987*	*% Long-Term Unemployed, January 1989*[a]	*Lone Parent Households, 1986*[b]	*% of Population Over Pension Age, 1987*[c]	*% of Households Lacking Inside WC, 1981*	*Gross Value Added in Manufacturing, 1988*[d]
Birmingham	–1.9	11.6	49.1	45.0	17.7	3.5	2,166
Bradford	–0.6	11.9	43.4	18.0	17.1	2.8	694
Coventry	–2.5	10.8	48.5	13.0	17.7	2.3	888
Manchester	–2.0	12.6	43.6	24.0	18.0	4.2	684
Leeds	–0.3	11.3	39.0	27.0	18.6	1.0	1,064
London							
Hackney	2.1	10.5	48.6	13.0	15.6	2.5	155
Haringey	–5.3	10.5	38.1	10.0	15.5	2.5	131
Tower Hamlets	10.0	11.7	48.8	8.0	15.9	2.8	185

SOURCE: Central Statistical Office (1989).
a. Out of total unemployment for the area.
b. In thousands.
c. Women over 60 years old, men over 65 years old.
d. Millions of British pounds.

controversy about the inadequacy of public funding for urban programs and drew attention to the question of how the inner cities could sustain social and improved life in run-down inner-city neighborhoods when resources were constrained.

Corporate Visions

These tensions affected relationships between the city council and community organizations. The riots, and the poll tax issue mentioned below, generated conflict between the council and community leaders that challenged the assumptions behind the modern version of Birmingham's civic gospel (Jacobs, 1986, 1988, 1992). Pride in the city required a workable partnership among local interest groups, the city council, and the private sector based on compromise and trust.

It was such trust that was sought by the influential national business employers group, the Confederation of British Industry (CBI), in its 1988 report, *Initiatives Beyond Charity.* The report was supported by central government and the private sector-led Business in the Community organization (the president of which was Prince Charles), which had a record of success in promoting public-private partnerships. The CBI (1988) was committed to forging a "shared vision" in the cities that would involve community organizations to build confidence in the cities and develop new structures for growth. The CBI supported initiatives in Birmingham and other cities that stressed the importance of local action, but with communities taking advantage of the assistance offered by national companies and, where appropriate, newly formed "business leadership teams."

Conflict and Community

With a Labour council committed to regenerating Birmingham by way of public-private developments in the central business district and elsewhere, there were inevitably allegations that the council was losing touch with the people at the grass roots by fostering the corporate vision. The tensions evident in connection with race-related issues, competition for central government urban program funding, and planning proposals affecting the future of the central business district and its traditional small traders all pointed to the problems involved in creating a shared vision.

On the positive side, community leaders generally adopted a moderate political orientation in their dealing with Tory and Labour politicians. This was evident even after the 1981 and 1985 riots, when there was a clear desire by ethnic minority leaders to maintain good relations with the local authority so that communities could recover from the devastation. However, there was no long-term guarantee that harmony could always be assured. This was especially so because run-down communities often felt left out of the internationalizing process, as social inequalities remained and as small traders experienced the adverse impact of large retail developments and the spatial reorganization of large parts of the CBD.

Centralization and Fiscal Pressures

In this context of social disorder and market change, there was a danger that the city council could be gradually transformed into a facilitating and regulatory body with a social welfare role for the deprived. It needed to ensure that it maintained a distinct decision-making role, but a crucial aspect of this was the council's retention of sufficient resources to fund and develop its activities.

However, its resource base was weak. The council maintained a politically moderate stance so as not to antagonize central government, but the Thatcher government's policies on municipal spending and the poll tax were deeply worrying to the local Labour party.[2] The city adopted a nonconfrontational approach, but local Conservatives continued to criticize the city council for its "high spending" policies. Also, the Thatcher government eliminated the top-tier West Midland Metropolitan County Council in the attack on the big "metro counties" that were abolished in 1986 (along with the Greater London Council) as part of the offensive against the "new urban left" (Stoker, 1988). In Birmingham, anti-poll tax rioters disrupted a council meeting and accused the Labour council of imposing the government's policy. Militants affiliated with the All Britain Anti-Poll Tax Federation alleged that the local Labour party had "sold out" on the tax issue. The city council was therefore politically ensnared between the government and the protesters.

In 1989-1990, Birmingham's net revenue spending in real inflation adjusted terms was down from the previous year (see Figure 4.1), with the trend likely to continue on a downward, or at best flat, path for 1991-1992 (Birmingham City Council, 1990, 1991). However, when

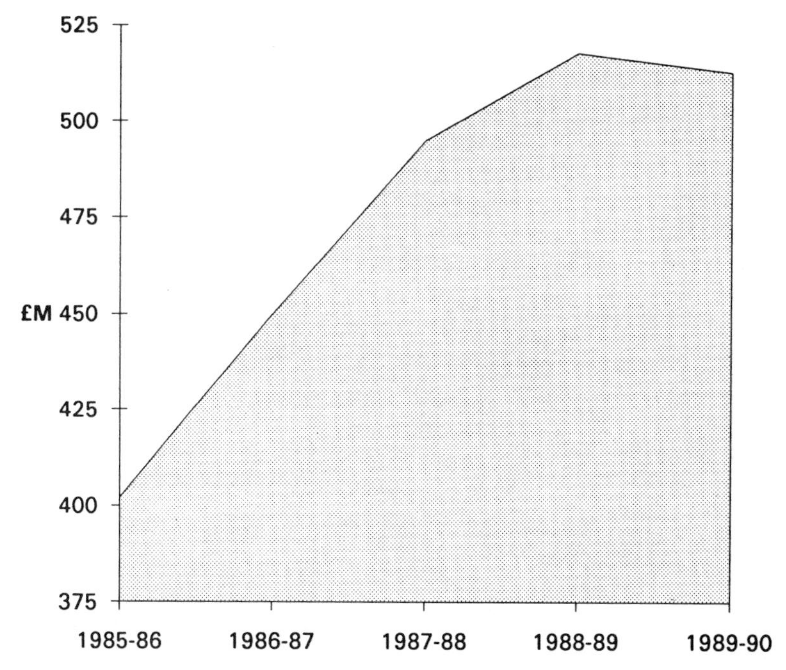

Figure 4.1. City Council Net Revenue Spending in Real Terms, From 1985-1986 to 1989-1990, at 1985-1986 Prices (figures rounded)

SOURCE: Adapted from Birmingham City Council (1990).

the West Midlands County Council was abolished, some services were transferred to the city. This caused a substantial rise in spending attributed to the city after 1986-1987 (see Table 4.2 and Figure 4.1) despite the use of special reserve funds to keep spending down. The lack of special funds in 1987-1988, plus increased levels of pay for council workers and capital items, contributed to a continued upward movement.[3]

The increase in city council net expenditure during the mid-1980s accompanied a real reduction in central government block grant funding that was reflected in the increases required to transfer the former county council services. As the rate of increase in the government's grant fell below the rate of inflation, the city faced upward pressure on local taxes and council service costs.[4]

In 1991, the council announced that its poll tax level would remain unchanged. In that event, the tax "freeze" actually enabled the tax rate to be reduced. The government subsidized the community charge following

Table 4.2 Birmingham City Council Five-Year Revenue Expenditure Financial Record

Item	*1985-1986*	*1986-1987*	*1987-1988*	*1988-1989*	*1989-1990*
Rates income[a]	236.0	311.6	335.2	350.4	376.9
Block grant	178.8	182.8	183.3	194.7	241.8
Total	414.8	494.4	518.5	545.1	618.7
Net expenditures of committees before contributions[b]	406.3	492.1	539.1	585.6	623.6
Deficit of committees before contributions	8.5	2.3	20.6	40.5	4.9
Net balance after contributions and other items	26.7	69.9	52.8	15.3	12.1

SOURCE: Birmingham City Council (1990, 1991).
a. Including domestic rate relief grant. The community charge and the unified business rate replaced the rates in 1990-1991. Estimated outputs for 1990-1991 are not shown here.
b. In terms of main service provisions (excluding certain "special services"), in 1989-1990, the council spent 52% of net spending on education, 3% on housing, 3% on economic development, 3% on urban renewal, 5% on environmental health, 16% on social services, 9% on leisure services, and 7% on technical services. The city council spent £226.6 million on capital schemes in 1989-1990. Estimates for net revenue expenditure for 1990-1991 were £776.5 million, and for 1991-1992, £862 million (not adjusted for inflation).

the announcement of a transfer of some of the burden from the poll tax bill to a national VAT tax bill for 1991-1992 (this did not affect business rates). The five-year financial position of the council is shown in Table 4.2, which indicates the changing position on block grants and the effects of the tight budgeting of a council keen to keep within central government guidelines.

Conclusion

Economic and political changes produced a very different corporate and political culture to that of the nineteenth century, when Birmingham's manufacturing base was more localized and largely British owned. Local political and business elites formally represented the church and industry. Now, they are more socially diverse and internally segmented, and the core decision makers have to link with the international corporate private sector. The "externalization" of power reduced local autonomy despite the council's successes in implementing innovative policies to strengthen the municipal role in economic development.

The Labour administration operated where the market-led and pro-foreign inward investment philosophy of the Thatcher era accelerated the restructuring of public-private sector relations and strengthened the role of corporate interests in shaping the destinies of major cities like Birmingham (Jacobs, 1992). This happened concurrently with the shift of power away from British local to central government (Davies, 1992; Stoker, 1988).

Notes

1. The main interests were in Birmingham Heartlands Ltd. (the east Birmingham urban regeneration initiative prior to the establishment of an urban development corporation in 1992), the East Birmingham Urban Development Agency (Birmingham City Council, 1988, 1989a, 1989b), Birmingham International Airport Ltd., the Birmingham Technology Group, Heartlands Development Services Ltd., Hyatt Regency Birmingham Ltd., and National Exhibition Centre Ltd.

2. The "poll tax" (the popular name for the community charge implemented in 1989 in Scotland and in 1990 in the rest of the country) replaced local "rates" or property taxes. The rates had been levied on individual householders. The poll tax was levied on all eligible adults in an attempt to make local councils more accountable to a wider number of people who used local services. The Major government decided to replace the poll tax with a council tax (to be implemented nationally in April 1993) as a partial return to property-based local taxation, but combined with an element of poll tax-style adult "head taxing" levied on at least two principal adult householders (with financial relief available for single householders). Many Thatcherite Conservatives saw the council tax as a retreat from the original objective of achieving local accountability through universal local taxation.

3. The 1989-1990 decline in real net revenue spending accounted for the removal from the overall figure of expenditure on education after the government's removal of the polytechnics from local authority jurisdiction. This produced a more realistic picture of spending in "mainstream" programs.

4. After 1988, the government eased its block grant provisions by allowing the council to offset the poll tax burden slightly.

References

Birmingham City Council. (1986). *Birmingham: The business city—1986 economic strategy review.* Birmingham: Author.

Birmingham City Council. (1988). *Birmingham Heartlands: Development strategy for East Birmingham, technical appendix.* Birmingham: Author.

Birmingham City Council. (1989a). *Birmingham Heartlands waterlinks development framework.* Birmingham: Author.
Birmingham City Council. (1989b). *Nechells development framework: Draft for consultation.* Birmingham: Author.
Birmingham City Council. (1990). *Annual report and accounts, 1989-90.* Birmingham: Author.
Birmingham City Council. (1991). *Budget, 1991-92.* Birmingham: Author.
Birmingham Inner City Partnership. (1989). *Programme, 1989-92.* Birmingham: Author.
Birmingham Post and Mail. (1991). *Municipal year book, 1991-92.* Birmingham: Author.
Briggs, A. (1963). *Victorian cities.* Harmondsworth: Penguin.
Central Statistical Office. (1989). *Regional trends 24.* London: Author.
Cole, G. D. H., & Postgate, R. (1966). *The common people, 1746-1946.* London: Methuen.
Confederation of British Industry. (1988). *Initiatives beyond charity: Report of the CBI Task Force on Business and Urban Regeneration.* London: Author.
Cowen, R. (1990). Urban design: Participation techniques. *Architect's Journal, 31,* 59-61.
Davies, H. (1992). Telling it like it is: The current balance of power between central and local government is unstable. *Municipal Review and AMA [Association of Metropolitan Authorities] News, 728,* 61.
Game, C., & Prior, D. (1991). The new partnership between Birmingham University and City Council. *Municipal Review and AMA [Association of Metropolitan Authorities] News, 723,* 197-198.
Gould, B. (1989). *A future for socialism.* London: Jonathan Cape.
Harding, A. (1990). Local autonomy and urban economic development policies: The recent UK experience in perspective. In D. S. King & J. Pierre (Eds.), *Challenges to local government* (pp. 79-100). London: Sage.
Holyoak, J. (1991). Waterside City. *Architects' Journal, 22,* 24-27.
Jacobs, B. D. (1986). *Black politics and urban crisis in Britain.* Cambridge: Cambridge University Press.
Jacobs, B. D. (1988). *Racism in Britain.* London: Christopher Helm.
Jacobs, B. D. (1992). *Fractured cities: Capitalism, community and empowerment in Britain and America.* London: Routledge.
King, D. S., & Pierre, J. (Eds.). (1990). *Challenges to local government.* London: Sage.
Knowles, R. (1989). Looking back to the future. *Municipal Review and AMA [Association of Metropolitan Authorities] News, 700,* 116-117.
Leppard, D. (1990, January 27). Birmingham's biggest: The Snow Hill development. *Estates Gazette* (Suppl.), pp. 80-86.
Longbottom, L. (1992). Promoting the city: Birmingham's public affairs department. *Municipal Review and AMA [Association of Metropolitan Authorities] News, 725,* 255.
Pipe, J. (1992, July 17). Caught in the storm. *Local Government Chronicle,* p. 21.
Rabeneck, A. (1990). The American invasion. *Architects' Journal, 28,* 36-57.
Rogers, R., & Fisher, M. (1992). *A new London.* Harmondsworth: Penguin.
Smith, A. (1991, May 24). Birmingham financial centre: Weathering the storm. *Investor's Chronicle,* pp. 53-58.
Stoker, G. (1988). *The politics of local government.* Hound Mills, UK: Macmillan Education.
Stringer, I. (1990, January 27). Offices booming into the 1990s. *Estates Gazette* (Suppl.), pp. 74-79.
Wray, I. (1988). Heart surgery in Birmingham. *Architects' Journal, 20,* 15.

5

Local Government in Poland

Political Failure and Economic Success

WISLA SURAZSKA

The most challenging task faced by the first post-Communist governments in Poland is that of setting in motion the mechanisms of both economic growth and social redistribution. Under the former regime, these two functions had largely been performed by a huge sector of the state-owned industry. The dismantling of this sector, although necessary by the criteria of economic rationality, has been tantamount to the elimination of the only existing mechanism of both industrial production and social protection.

It would seem only too obvious that the role of local government in managing this painful transition should be substantial. In the brief account below, only the contours of emerging local governments in Poland are presented. What place the local government will take in the development of both democracy and economy in Poland will depend on a complex set of circumstances. In particular, three factors will be

AUTHOR'S NOTE: Research reported in this chapter was financed by the Norwegian Council for Sciences and Humanities. Primary sources included 14 interviews with Polish representatives, state officials, local officials, members of the Government Commission on Administrative Reform and the Parliamentary Commission on Local Government, and advisers to both commissions, and archival materials, including accounts in the Polish press, statistical annuals, legal acts, and other government documents.

discussed in this study: (a) the ideas underpinning central-local relations in Poland, (b) the prospective development of the territorial system, and (c) the outcomes of economic reforms already under way. The impact of these three factors on the development of local government in Poland will be summarized in the concluding section.

The Emerging Contours of Local Government in Poland

Local Elections and the Role of Citizen's Committees

Local elections in Central Europe quickly followed the first democratic parliamentary elections.[1] But in this context Poland was an odd case. In spring 1990, when most of the countries of the former Soviet bloc had already acquired fully democratic representation, the Solidarity government was left with a parliamentary majority selected by a nondemocratic process agreed on early in 1989 at the Round Table. Although demands for new parliamentary elections rose in Poland from various corners, Premier Mazowiecki considered it unwise to throw the country into an electoral campaign in the midst of a painful economic transition. Local elections were called instead, as a compromise solution.

In May 1990 the Poles elected some 52,000 councillors in the rural townships and cities, each community with the status of a basic administrative unit—*gmina*—and designated as a site of local government. The electoral law of March 19, 1990, provided for proportional representation in constituencies of more than 40,000 and majority voting in the smaller ones. With only two months for the selection of candidates and electoral campaigns, the entire event appeared a little superficial. There were problems with the recruitment of sufficient numbers of reasonable candidates, especially in the rural areas. By mid-May, the main contenders, the Citizen's Committees, were still short 30% of the candidates, especially in the rural areas.

Only 42% of the electorate turned out for the May elections. The number was low considering that these were the first free elections for local government in postwar Poland.[2] One year later, in May 1991, the turnout in local by-elections sank to 13%. In one *gmina* only one voter showed up—he was a candidate, but he did not vote for himself. Clearly, something went wrong at the outset of the Polish experiment with local democracy.

A standard explanation of such failure—the lack of grassroots political organizations in the localities—did not apply in this case. In fact, the main contender in the local elections, the Citizen's Committees, represented the grass-roots movement: Thanks to the popular mobilization ignited by and channeled through Citizen's Committees, Solidarity won all the seats designated for contest in the national election of June 1989.[3] Thus the Citizen's Committees had accumulated a great deal of local energy, especially in the larger cities. Their role under the first Solidarity government of Premier Mazowiecki, however, was ambiguous until the spring 1990 local elections. There were some 1,200 organizations putting up candidates, but the only serious challengers to Citizen's Committees and Solidarity appeared to be independent candidates. The results of these local elections depended on whether the local industrial Solidarity cooperated or competed with the local Citizen's Committee. When the two ran a single list they always received a large majority, but when they competed with each other they both suffered, to the benefit of the third parties. The resulting composition of local councils is shown in Table 5.1.

Citizen's Committees and other Solidarity-related groupings were especially successful in large cities, whereas in the rural areas independents prevailed. Among political parties, the most successful appeared to be the Polish Peasant party (3.2%), recently transformed from the former Communist-licensed Peasant Union.

The Local Executive Board: Strengths and Limitations

In these newly elected local governments, the local executive consists of a board of up to five members, headed by a mayor (Table 5.2). The executive board is elected by the respective local council, but the mayor and deputy do not need to be members of the council. There are elements of a strong executive, as the mayor heads the local administration and also bears responsibility for the development and implementation of political and economic strategies. Some constraints on the executive powers may come, however, from the council, which can recall the mayor at any time by a majority vote. In fact, the efficiency of the executive largely depends on the terms of cooperation between the mayor and chairman of the council. There have been cases in which strife between the two has paralyzed local authorities. In such cases the central government has the right to appoint a commissioner to run local matters. Perhaps this is one of the reasons the Council of Ministers opposes attempts to strengthen the local executive by merging the positions of mayor and chair of the council (member of the Government

Table 5.1 Results of Local Elections in Poland, May 1990

Affiliation of Councillors	*% of Seats*
Citizen's Committees and other organizations connected to Solidarity	45
Six national political parties and five connected coalitions	10
Social organizations (cultural, professional, regional, etc.)	5
Independents	39
Others	1

SOURCE: Author's calculations from GUS (1990).

Commission for the Reform of Public Administration, personal communication).

Further limitations on the powers of local executives have resulted from the creation of an additional subprovincial tier—*rejony*. It became necessary to introduce these supplementary units because of serious flaws in the territorial system left behind by the previous regime (see the discussion in the following section). *Rejony* were designed as an "administrative police," or the *voivode*'s long arm stretching into the localities; they took over several functions formerly reserved for local governments. As a result, the latter's authority has been undermined by central interference into local affairs.

Two Competing Patterns of Local Government

The fact that fully democratic local elections preceded the national ones by more than a year had some impact on the further development of central-local relations in Poland. First of all, it enticed more political activism from local authorities than the central government was prepared to accept. For example, the National Convention of Local Governments, an extraconstitutional body created soon after the elections, claimed itself (unofficially) to be "the sole democratic assembly in the country" (*Tygodnik Solidarnosc,* August 24, 1990). Second, at the time of the communal reforms, basic constitutional issues remained unresolved. Polish laws on local government were prepared at a time when the Sejm was already clogged with an avalanche of bills on the ongoing economic transition, hence the lack of clarity and numerous gaps in the statutes concerning communal reform.[4]

Table 5.2 The Polish Territorial System in 1992

Unit	*Type of Administration*
Subnational units, 49 *wojewodztwa* (population average 292,000)	*wojewoda* (prefect), centrally appointed official for all-purpose administration; *sejmik,* representation of local governments, advisory powers
Subprovincial units, 267 *rejony*	consolidated office of services subordinated to ministries with *rejon* chief appointed by *wojewoda*
Basic units, 2,459 *gminy*	
Types of *gminy* and % of total population represented	sites of elected councils
1,618 rural *gminy* (28%)	*wojt,* elected by council
535 rural-town *gminy* (21%)	*burmistrz,* elected by council
299 town *gminy* (46%)	*prezydent,* elected by council
7 districts of Warsaw (5%)	*burmistrz,* elected by council

SOURCE: GUS (1992).

The legislative intent was to hand over decisions on local matters to local communities. But such a general statement did not provide enough guidance for specific problems of division of powers and institutional solutions. From the start, the institutions and assets of local government received legal recognition as distinct from those of the state. This put an end to the Communist principle of uniform state power and property. The details of this separation had not been well elaborated, however. It seemed as if two different or even contradictory historical patterns were intermixed in the project. The first came from the principle of unity of command deriving from traditions prior to the war; the second stressed a natural right of localities to self-government and a consequent withdrawal of the state from local matters.

The first pattern stems from the prewar Second Polish Republic (1919-1939), a unitary state shaped after the model of the French Third Republic. This was a blueprint for a "government by assembly," with a rather unstable cabinet at the center and a firm central executive in the provinces. When the cabinets of the Second Polish Republic were collapsing one after another, continuity and stability in the provinces was preserved by the *voivodes,* the Polish version of the French prefects—that is, centrally appointed officials who supervised all public services operating in the provinces. This reflects the corporatist conception underlying prewar local government in Poland: Local government had the status of an interest organization, with the rights and duties that came with it. But as the functions of local government overlap with those of the state (taxation, lawmaking, and so on), the central government has a right to supervise its activities more closely than those of other organizations.[5] Nevertheless, the Second Republic's structures of central administration ended at the subprovincial tier of the *powiat,* where they existed in a delicate balance with the elected assembly (*sejmik*) and the centrally appointed *starosta* (subprefect), who headed the assembly's executive. The rural townships and the cities (with the exception of Warsaw) elected their own executives and were fairly independent.

The second approach stems from the Roman Catholic doctrine of subsidiarity as formulated in the Papal Encyclical of 1892 (*De Rerum Novarum*). At that time the church sought to uphold the right of communities to self-govern in the face of the rising powers of European nation-states. Thus the doctrine of subsidiarity recommended the performance of public tasks at the lowest possible level. It should be stressed, however, that the Catholic church in Poland did not show much interest in the communal reforms of 1990.

A secular and more radical version of a communitarian option was developed in the course of the Weimar constitutional debate. The opponents of the traditional presumption of state sovereignty proposed a "new epoch state" based on self-governing communities as an alternative (Schmitt, 1985, pp. 22-27). The embodiment of the new epoch state at that time was the looming system of soviets in Russia, which merged legislative, executive, and judicial powers within directly elected councils. Despite the obvious failure of the Soviet experiment, the idea has persisted in New Left approaches that stress the advantages of communal self-organization and participation over centralized state bureaucracy.

Central-Local Relations and Intermediary Groups

The emerging center-periphery relationships in Poland can be best understood in terms of the two competitive designs, corporatist versus communitarian, described above. The centralized, unitary model was defended as the only way to manage the process of transition. Furthermore, the need to strengthen state institutions after the collapse of the Communist regime was emphasized. The supporters of a communitarian model, in turn, raised the question of democracy and the need to strengthen civil society after years of dictatorship. The weakening of a centralized state, they argued, is an opportunity to start building it from below.

Apparently, the prewar pattern of dual (state and local) administration now has resumed. The provincial (*wojewodztwo*) authority belongs to the centrally appointed *voivode,* but there is also a representative body of all local councils in the province (*sejmik*). The division of powers between the *sejmik* and the *voivode* is not quite clear, however. The latter can suspend a particular ruling of a local council, but the *sejmik* also can disapprove of the *voivode* or his decisions, although the consequences of such moves are not prescribed by the law. There have been cases of *voivodes* staying on after the *sejmik* has voted against them. Further, the *sejmik*'s collegium has some powers of arbitration independent of the regular network of administrative courts; this reflects somewhat the communitarian ideal of merged powers.

Another element of the communitarian model has been the open-ended provision for the powers of local government: Local governments are entitled to exercise authority over all issues that have not been reserved for the state. This provision has been undermined, however, by an executive ruling in which some 300 tasks of local government are carefully listed (Kulesza, 1991). Further limitations ensued from ministerial decrees that took away certain vital local government powers, such as the licensing of construction developments, land sales, and water and communication enterprises.

The interests of local government clearly have not been secured either by means of constitutional provisions or by other legal-institutional arrangements. Nevertheless, certain political levers to influence central decisions have been developed, first by the Citizen's Committees and, upon their dissolution, by the regional and national associations of local government. The Citizen's Committees had their period of glory when, in the absence of political parties, they enjoyed political clout both in the center and in the localities. Parliamentarians depended on the committees for reelection and, in many cases, chairs of the committees headed local executives as mayors. In such cases, the influence of the committees overshadowed the city councils. For example, the mayor of Przemysl considered the local Citizen's Committee (which he was heading) a necessary check on some councillors who might play dirty tricks behind his back. He complained about the excessive influence of the former Communists who found their way to the city council (personal communication).

Citizen's Committees earned this high moral standing by leading crusades against local business groups connected with the former Communist apparatus. The committees were expected to monitor these activities and to deprive such groups of ill-gained profits. Paradoxically, in this way many Citizen's Committees took upon themselves the former party apparatus's role as the brokers and supervisors of local interests. In summer 1991, the committees split into a multitude of political parties to contest the national elections. As a result, the ideological stance of local governments subsided and the role of defending local interests was taken over by the associations of local government.

Three national associations represent local governments in Poland. The National Convention of Local Governments has been created as the emanation of the provincial assemblies of local councils. Its formal-legal status is unclear, but it has acquired an advisory status in matters of central subsidies for local governments. Cities set up their own national conventions to promote their interests by forging international business links. There is also a separate Association of Polish Communes representing chiefly the rural areas, though it has few political levers of its own. Nevertheless, it enjoys some measure of support from the Ministry of Finance, which plays a skillful game among various representatives of local governments when it comes to the distribution of central grants (adviser to the Parliamentary Commission for Local Government, personal communication).

In addition, there are numerous regional associations of local governments that cover more or less the traditional regions. They started as cultural and social organizations, only to become increasingly political in character in recent years. The main initiators have been the mayors of the principal cities, the *voivodes* of the relevant provinces, and the voluntary associations of local government, which often cover the territory of the old *wojewodztwa* that had been cut into pieces in 1975. They take various forms, beginning with cooperation of the present provinces initiated by the *voivodes* to associations of local self-governments engineered by ethnic minorities, as is the case in Silesia.[6]

The main objective of these regional organizations is to gain recognition in the administrative structure, but demands for devolution of legislative powers are present as well. Such associations were allowed by statute, although the legislative intent was to encourage economic cooperation between local governments rather than political activism. In response, the Council of Ministers Office has stressed the provisional character of the country's administrative division and the exclusive right of the government to decide about its future changes ("O Reformie Administracji," 1990).

The Territorial System and Its Impact on the Power Structure

One of the most daunting legacies of the Communist regime has been the territorial division of the country. Merging and subdividing the administrative units was a constant practice of the Communist leadership as it tried to prevent too close an integration of local and provincial elites. The most radical territorial reorganization happened in Poland in 1975, when all the established administrative units were dismantled and entirely new divisions were created. The provinces (*wojewodztwa*) were reduced to about one-third of their former size. Only a small proportion of the former provincial boundaries, many of them historic, was preserved. Furthermore, the intermediary tier of *powiat* was abolished and new subprovincial units—*gminy*—were created instead.[7] *Powiaty* were not only the cornerstones of the prewar local government, they were also the ancient centers of local trade and politics. They developed organically throughout the centuries in such a way as to make it possible for all the inhabitants to reach the principal town and to return home in the course of one day by horse cart. The local networks of communication, transportation, and supply had developed on the *powiat* basis. Not only were these ancient units dissolved in 1975, but some 30% of them were split between two or more new provinces.[8]

The consequences of this reorganization were far-reaching. Generally, the country became unmanageable, both from the center and from within the new provinces. The new provinces consist of some 50 to 70 *gminy,* many with little means of communication with their new centers. The new provincial capitals lacked sufficient facilities and personnel for the all-purpose management typical of the Communist party administration. As a consequence, some 200 additional divisions for special purposes had to be introduced, all of them subordinated directly to ministerial branches in the center. Thus the regime became overcentralized and at the same time fragmented along functional lines. This unwieldy territorial structure was in place until recently (see Table 5.2).

The new administrative divisions became the single most persistent issue of grassroots grievances haunting the government of Jaruzelski and, later, each of the four subsequent democratic governments. There have been numerous press articles on the absurdities of this division. Thousands of petitions for change poured first into the Communist Party Central Committee, then to the Office of the Council of Ministers, which is responsible for the corresponding decisions. The newly elected local governments tend to ally themselves, whenever possible, with their former provincial centers rather than with the new ones. Functional fragmentation of the center acquired a powerful boost after the collapse of the Communist regime. The ministries became all-powerful, their territorial agencies multiplied, and their policies were difficult to control or even to coordinate, not only in the provinces but also in the center (Stepien, 1990, p. 7). Some regional associations of local governments have pointed to the present system's low levels of administrative performance in their claims for devolution of central powers.

Thus the central government found itself on the horns of a dilemma: either to give in to the pressure and accommodate local and regional aspirations, which meant the resumption of the traditional regions, or to keep the obsolete territorial system, which at least had the virtue of keeping in check subnational elites. A possible return to the old administrative division was seriously considered by the Mazowiecki government, but it was finally rejected as too costly and as posing too much risk to the country's precarious stability. The partial solution adopted was the restoration of some of the former *powiaty* under the new name of *rejony*.

Only a little more than half of the former *powiaty* were restored, and there was much agitation and rivalry among the townships for the acquisition of the much-desired *rejon* status. The resumption of the old *powiaty* as the secondary units of local governments has been pursued in the center by a small but influential all-national party, the Liberal-Democratic Congress.[9] Their program suggests not only the strengthening of local government but also deconcentration reforms in state administration that would strengthen *voivodes* by handing over to them many of the functions performed by the ministries.[10]

Nevertheless, the new provinces receive a powerful reinforcement in the Senat, the upper chamber of the Polish Parliament. Two senators are elected from each *wojewodztwo*. Because the population of *wojewodztwa* varies from 30,000 to 4 million, underpopulated regions are over-represented.[11] This arrangement favors the poorest regions and is likely to remain under the new constitution.

Local Finances

In this period of transition, both the central and local governments have been developing their own structures and resources in the presence of a growing fiscal crisis. The crisis resulted mainly from the collapse of state-owned industry, which was the prime source of state revenue at the central as well as the local level. It will be several years before private business becomes a substantial source of public revenue, and more time will be needed to establish an effective tax collection system. In this respect, local government seems to be more effective than the central one. Indeed, the financial achievements of local governments have constituted one of very few success stories of the Polish transition process.

This appears even more remarkable in the context of the initial circumstances. There used to be a close link between the state industrial enterprises and the maintenance of local infrastructure. Economic and political dominance of the workplace over the lives of individuals and communities was an important characteristic of the Communist regime. One ideal was to link services for the population with the people's performance as state employees. Therefore, the withdrawal of state subsidies to industry has had a secondary effect in the depletion of local services. Privatization will not solve this problem, because one cannot expect new owners to maintain sport centers, kindergartens, or water supplies to the localities, as was the customary practice of the state enterprises.

Thus local government has been burdened with the responsibility for the aftermath of the collapse of the centrally planned economy. Local councils must deal with the maintenance and development of local infrastructure as well as with the social and economic consequences of bankruptcies of big industrial enterprises located in their areas. The question is whether they have been properly equipped for these tasks. Local governments have been given property rights to their assets as well as limited rights of taxation, with the upper limits set by the Ministry of Finance.[12] Furthermore, some revenues were expected to flow from enterprises in the possession of local governments. Shares in some of the state taxes and central equalization grants were to supplement local budgets.

Before the new law on local finances came into effect in January 1991, more than half of local expenditures were financed by the center, mostly in the form of earmarked funds and emergency subventions. The central budget was balanced in 1990, a major achievement of the Mazowiecki government, but a large part of the central deficit was

Table 5.3 Dynamics of Local and Central Budgets in Poland, 1991

	Revenues Minus Expenditures	
	Central Budget	*Local Budgets*
First quarter	−6.127	3.555
Second quarter	−13.150	5.811
Third quarter	−22.377	6.426
Fourth quarter	−30.973	5.244

SOURCE: Ministry of Finance (1992, p. 96).
NOTE: Figures are in billions Zl; 1 billion Zl = $670,000. Total amount of local expenditures in 1991 was 45 billion Zl, accounting for 18% of central government expenditures.

transferred to local governments in the form of delegation of additional responsibilities (some health and educational services), often without adequate funding for their maintenance. This tendency persisted in 1991, when subsidies to local governments were cut by 11% as a result of a growing central deficit. Nevertheless, 97% of all local budgets in 1991, the first year of local independent budgets, ended with a balance and in many cases with a surplus.[13]

Where did the surpluses come from? It is difficult to assess how much of this can be credited to the activity of particular councils versus the role of accidental factors, such as prior distribution of vital assets or the existence of active business in the locality. Generally, there are three explanations for surpluses in local budgets. First, local governments acquired a lot of property, some of which, especially the real estate, they were allowed to sell. This was only a one-time enrichment, not to be expected in the future. Exact data about the size of this particular source of revenue are unavailable because they are lumped together with "own nontax revenues" (with earnings from rentals of communal property, local fees and charges, incomes brought by communal services, and so on). Nevertheless, a general tendency in local government strategies was to rent rather than to sell property (see the following section on communalization of state property).

The second explanation for local budget surpluses is the gross underestimation of tax revenues, especially those coming from small private businesses; these were almost triple the expected amount. This might be credited to the effectiveness of the local tax collection system but also to the greater overall expansion of small businesses. Whatever the case, the gains from the private sector offset the losses resulting from

Table 5.4 Performance/Plan Ratio of Local Revenues and Expenditures in Poland, Fiscal Year 1991

	Performance/Plan Ratio
Total revenues	130.8
Own revenues	188.8
Participation in central taxes	91.7
From state industry	40.5
From salaries	92.3
From private business	276.5
Central grant (general)	98.7
Expenditures	122.8

SOURCE: Ministry of Finance (1992, p. 96).

the lack of revenues expected from state industry, something that the center has yet to achieve.

A third explanation comes from both lack of experience and overly cautious planning by local councils, particularly urban governments (Grobelny, 1991, p. 15). Examination of the first independent urban budgets for fiscal year 1991 reveals that they were designed according to their predicted financial capacity rather than the real needs of the localities. This can be inferred from the lack of correlation between the expenditures per capita and the number of inhabitants in the cities. Such correlation can be expected for two reasons: (a) the well-known rule that financial needs grow faster than the number of inhabitants because of certain barriers in the technical efficiency of infrastructure, and (b) there was an additional central grant for the cities of more than 100,000. Thus the lack of correlation between the number of inhabitants and expenditures per capita testifies indirectly to the self-imposed financial discipline of urban councils.

Generally, local expenditures in the first independent budgets included a larger proportion of investments (25%) than found in the central budget, even after allowance is made for such additional central spending as social insurance and the servicing of foreign debts. Also, the proportion of subsidies that went to education and culture in local budgets exceeded those made by the center. Most of the local investments went into communal infrastructure and the construction of housing. Another substantial local expenditure item has been administration, especially salaries. As soon as local governments were allowed to construct their budgets independently, the salaries of local administrators rose considerably. As a result, local councils are able to compete for staff with the state administration, which has become virtually starved of the best specialists because of its rigid wage schemes.

Central (general) grants to local government amounted to 14% of all local revenues, but there was a marked difference between their role in the rural areas (31%) compared with towns and cities (9.2%). Generally, the final balance of local budgets did not correlate with the size of the central grant included in the revenues. This indicates certain inadequacies in the distribution of central subsidies. Even though the central subvention was calculated to account for prospective local revenues, the per capita expenditures in rural areas were still about 65% of those in towns and cities. The differences within urban areas have been even larger, mostly corresponding to regional differentiation. The major regional divide is the traditional one, into the poor East and rich West. The largest interregional differentiation of local revenues per capita is greater than 200% (Ministry of Finance, 1992, p. 101). Some analysts warn that these differences are threatening the integrity of the state (Malecki, Sobiech, & Refermat, 1991, p. 22).

Local Responses to Economic Reforms

Communalization of state property has been part of the communal reforms aimed at giving local government property rights to local enterprises and equipment necessary to maintain and develop local infrastructure and services. The relevant decisions have been made by *voivodes,* who were expected to represent central state interests. In fact, the allegiance of *voivodes* has often switched to local interests, which resulted in a number of revised decisions by the center.

It should be kept in mind that the parallel process of privatization of state property has already been under way as a major item of economic reform. The decision about whether a given enterprise is to be sold on the market or handed over to local authorities (who then may lease or sell it themselves) bears some financial consequences for the direction in which the cash will flow. The central budget had been constructed on the prediction of substantial income from the sale of state industries.

These targets could not be met, however, because large industrial enterprises were difficult to sell. The smaller and more profitable ones were in turn claimed by local governments, with the complicity of some *voivodes.*

The Ministry of the Transformation of State Property expressed some misgivings about communalization schemes that may hamper the privatization process. They pointed out that local governments tend to keep their property and lease it instead of selling. The result might be that the Polish economy will become communalized instead of privatized. Further, the center objects to the inclusion of local services in local budgets, because in this way it loses a substantial share of payroll taxes. Thus the center's recommendation to local authorities is to contract with private services instead of running them (adviser to the Parliamentary Commission on Local Government, personal communication).

The involvement of local government in profit-making activities has also been criticized by the Anti-Monopoly Commission. It has argued that, under the circumstances, the monopolistic position of local government may undermine the development of private business in some sectors. For several reasons, local councils have been slow to sell enterprises handed over to them. First, the market has been unstable and they are anxious not to undervalue their property. Second, domestic capital has been rather scarce, and property sales to foreigners would have unpleasant political repercussions. Finally, time and experience are needed before a general strategy of generating income for local spending is in place: Is it to be taxation of private business or profits from the council's own economic activity?

The rural councils were hoping for the land left behind by the bankrupt state farms, but the central government set up a specialized Agency of Agricultural Property to deal with these problems. Activities of this agency deserve separate study, especially because the problems of bankrupt state farms have been commonplace in the rural areas of all post-Communist countries. Suffice it to say that in Poland these problems have not been handled adequately and have caused many conflicts and damage to property as well as increasing unemployment in the rural areas. One of the reasons has been the lack of communication and sometimes even open discord between the Agency of Agricultural Property and local authorities. Whether the latter are better suited to handle these problems is an open question and an important one.

Conclusions

Which of the two frameworks of central-local relations outlined in the beginning of this chapter will succeed—the corporatist one, typical of a unitary, centralized regime, or the communitarian version, based on the assumption of a natural right to local self-government? In fact, neither has materialized so far in practice. There is a persistent tension between the center's aspiration for control and its capacity to maintain control. A centralized unitary regime can be achieved only with a long-term, costly strategy. The main components of such a strategy, as employed by the European nation-states, include (a) the development of a welfare state with centrally controlled redistribution of resources, (b) the existence of professional administrative personnel with a homogeneous ethos and unambiguous allegiance to the center, and (c) the stability of the party system, which would allow for political consolidation of the center (Rhodes & Wright, 1987, p. 18).

None of the conditions above has been met by the Polish government. First of all, there is not much in the way of resources that can be centrally distributed. The budget deficit has already exceeded the barrier of 5% set by the International Monetary Fund as the condition of financial assistance. Second, the party system is in turmoil; the 29 parties that found themselves in the Sejm as a result of the unlimited proportionality in the electoral laws have been multiplying still further, rendering the center politically ineffective. Finally, nothing similar to the Western European "administrative class," with professional training and nonpartisan status, exists in Poland.

A good example is the perennial failure of the center to acquire some measure of stability in the administration of provinces. Each time the government changes, the *voivodes* are vulnerable, because their positions belong to the pool of spoils shared by the subsequent governing coalitions (the fourth one in the past three years). The only way for a *voivode* to remain is to acquire local support and expect that the provincial gathering of local councils (*sejmik*) would approve of nobody designated by the center. This instability, of course, gives leverage to local government.

Does this mean that local government in Poland is going in the direction of the communitarian model? Not quite. There was a short period of grassroots political activism in the Polish localities in the form of the Citizen's Committees, but the results were not very encouraging. The ideological-moral approach held by many of the committees often provoked conflicts with the local councils and local business communities. Although local shopkeepers often have some connections with the former Communist apparatus, they are the only shopkeepers in the vicinity and thus represent a real economic force that no local government can afford to disregard. The disappearance of the Citizen's Committees made local politics less righteous but more effective.

The degree of self-government largely depends on local financial autonomy. In this respect, the urban councils have been much more successful than the rural ones, the latter being more dependent on central grants. In fact, the main challenge for both central and local governments is coming from the "structural poverty" already forming in rural areas dominated by economically unviable state farms and industry. There is little local activism in such precincts, and both inhabitants and their elected councils look to the center for help. The passivity and the resulting poverty of such areas cannot be cured by means of self-government and participation because of the lack of communal bounds among the inhabitants.[14]

The richer urban areas, in turn, strive for more independence from the center, but their goal is not typically communitarian. City dwellers are not very keen on building up a community life by participation in local decision making and the like. What they are interested in is professional delivery of the services they need and a reasonable level of control on public spending. There have already been growing tensions between the specific needs of urban administration and communitarian elements of local government included in the statutes. The city of Warsaw, for example, became excessively fragmented under the slogans of communal participation.

It is becoming quite clear that centrally directed economic transition has its limits. Capitalism itself is a grassroots phenomenon that can be encouraged by macroeconomic policy, but only up to a point. The role of local government in both stimulating economic growth and providing protection against its side effects has been increasing. Local governments proved able not only to provide local services but also to support the limping state health and education services. In addition, local governments have done all this with their budgets balanced.

Thus the question is not whether local government is able to contribute to economic transition, but whether it will be allowed to do so. Part of the answer is included in the prospective changes of the territorial system. Some of the historic regions in Poland have already shown potential for economic development that much exceeds the centrally directed program of transition. Thus a choice for the center is either to slow down economic development in order to maintain a unitary structure or to give away some powers to the regions and accelerate the transition. So far the first option has been vigorously defended by all four democratic cabinets, even though the price is a kind of administrative vacuum in the provinces.

Notes

1. In Hungary, elections were held in September 1990; in the Czecho-Slovak Republic, in November 1990; and in Poland, in May-June 1990.

2. The corresponding numbers in East Germany and Czechoslovakia were about 75%; in Hungary, notorious for its low electoral turnout, it was still 51%.

3. The first Citizen's Committee was created as an advisory body to Lech Walesa on December 18, 1988, in anticipation of the Round Table negotiations the following spring. In June 1989, the committee became responsible for electoral campaigns on behalf of Solidarity. The 160 participants in the committee were invited personally by Walesa.

4. These laws included those of March 8, 1990 (on local government, Dziennik Ustaw No. 6); March 22, 1990 (on the territorial organs of central administration, Dziennik Ustaw No. 21); May 10, 1990 (the property law for local government, Dziennik Ustaw No. 32); and Ustawa Kompetencyjna (the law on competence), dividing powers between *voivodes* and local governments (June 9, 1991).

5. Nonetheless, local governments of the Second Republic conducted lively and independent economic activities. Their budgets amounted to about 30% of the national one, their property was legally defined and protected, and they were entitled to draw foreign loans. State subventions amounted to merely 7.5% of local budgets (*Statistical Annual,* 1939, p. 397).

6. I have completed research on this issue, and I am now in the process of analyzing data.

7. Abolished in 1954, *gminy* were reinstated in 1973, but they were much smaller in size and their number kept changing.

8. This figure comes from my own calculations.

9. I am grateful to Representative Lech Mazewski, who made available materials and meetings on the subject.

10. Local election (instead of central appointment) of a *voivode* is still another matter and is less emphasized at the moment.

11. According to my calculations, 21% of the population living in the smallest provinces is represented by 38% of the senators.

12. More exactly, following laws passed December 14, 1990, and January 12, 1991, the revenues of local government consist of the following taxes: agricultural tax, property tax, transportation tax (vehicles), inheritance and donation tax, taxes on business of natural persons, and dog tax. Further revenues come from fees for administrative services, income from the assets in possession of local government, the share in central taxes (wage and income taxes), and central subsidies, general as well as targeted.

13. A new system of local finances was introduced January 1, 1991, but it is due to change after the system of taxation is reformed (the introduction of value-added tax).

14. Electoral results have already shown that these sites have politically explosive potential; most votes for political extremism come from these areas.

References

Grobelny, R. (1991). *Analiza sytuacji finansowej miast polskich* [Analysis of the finances of Polish cities]. Poznan: Zwiazek Miast Polskich [Association of Polish Cities].

GUS [Main Statistical Office]. (1990). *Statystyka Wyborow do Rad Gmin 27 May 1990* [Statistics of elections to local councils 27 May 1990]. Warsaw: Author.

GUS [Main Statistical Office]. (1992). *Powierzchnia i ludnosc w przekroju terytorialnym* [Space and population in territorial perspective]. Warsaw: Author.

Kulesza, M. (1991). Zagrozenia reformy; W sprawie zadan i kompetencji samorzadu terytorialnego [Reform at risk; Concerning the tasks and competences of local government]. *Samorzad Terytorialny, 1-2,* 86-90.

Malecki, M., Sobiech, J., & Refermat, P. (1991). *Z badan nad finansami gmin* [Study on the finances of *gminas*]. Poznan: Krajowy Instytut Badan Samorzadowych [National Institute for Local Government Research].

Ministry of Finance. (1992). Finanse Gmin [Finances of *gminas*]. In *Annual report.* Warsaw: Author.

O reformie administracji. (1990, June 17). *Zycie Gospodarcze.*

Rhodes, R. A., & Wright, V. (1987). Tensions in the territorial politics of Western Europe. *West European Politics, 4.*

Schmitt, C. (1985). *Political theology: Four chapters on the concept of sovereignty.* Cambridge: MIT Press.

Stepien, J. (1990). Kilka ostroznych krokow [A few careful steps]. *Samorzad Terytorialny, 1,* 3-12.

6

Market as Pressure or Market as Possibility?

The Framework of Local Economic Development in Hungary

GÁBOR PÉTERI

There are two interrelated processes in Eastern and Central European countries in recent years: decentralization or deconcentration in public administration, initiated by political changes in the late 1980s; and the emergence of a market economy, characterized by new forms of privatization in the formerly state-dominated system. As the first symbolic steps from a centralized state toward local autonomy-based governance, legal, administrative, and fiscal transformations occurred rather quickly, but economic changes have been slower. Foreign debt burdens and the collapse of Eastern European markets forced structural adjustments, although the privatization process has been slower than expected.

The case of Hungary illustrates two basic economic development issues facing the new local governments. The former Soviet-type councils, and now the municipalities, are the basic units of general government finances, but their roles are defined differently now. The shift from a "self-accounting" organ in a planned economy toward a self-financing, modern, local government opens up a great variety of political options and fiscal structures. Public finances, including local budgets, are under market *pressure*: Municipalities are forced to implement local economic development strategies, otherwise they cannot meet local financial and political requirements. On the other hand, the emerging market offers new policy options for local governments: The market offers the *possibility* of decentralized economic coordination in local development and the use of alternative service delivery methods. The question is how local governments will build on the market and what types of policies they will formulate.

Local Government Finances in Hungary

The planned socialist economy and its corresponding administrative structure were built on direct contacts of producers and consumers (Kornai, 1982). In this "resource-bounded" economy, lacking a price mechanism, decentralized ("vegetative") regulation and planning were based on quantitative figures. The vertical flow of information (reporting needs of customers, stock, capacity) was amplified by horizontal links. Socialist or state-owned property was, of course, an additional feature of the economic structure. Redistribution mechanisms and hierarchical separation of political and administrative institutions helped to stabilize this system.

But the idea of local "self-accounting" existed even in strongly centralized economies. In the former Soviet Union, the autonomy of the republics was to be increased through self-accountability (*hozraschot*). In the federative system of Yugoslavia, following centralized rules, municipalities and regions tried to define public revenues, expenditures, and economic plans (level of production, taxes, contributions, wages, and so on) at the local level. The goal of self-financing, as the basis of local autonomy, meant that each municipality should rely on the local economy inside its administrative borders or metropolitan area.[1] This notion of local government finances based solely on local "revenue-producing capacity" (Vági, 1991), however, was criticized by those emphasizing the differentiated economic structures of urban and rural areas, high rates of daily commuting, and so on. It was argued that, given changes in technology and management, the closed urban economy had disappeared and administrative structures could no longer be based on this assumption (Wiener, 1982). Overall, these local self-financing activities were hypocritical and futile efforts, given the central allocation of resources and investments in production and the public sector. Nevertheless, the long history of reform efforts based on the links between enterprises and municipalities serves as a fiscal *precondition* for local self-government in formerly socialist countries.

Regardless of its political and institutional characteristics, the role and function of the state can be measured by government expenditures. In the long run, the proportion of government expenditures relative to gross domestic product is declining, but in Hungary during the past few years—even in the period of rapid political transition—it is unchanged and, in international comparison, relatively high (average in the period 1986-1991 was 65%). This indicates that the connection between central and local governments and the economy remains strong and that the direct and indirect influence of the state is still significant.[2] But behind these general data, there is a shift from direct allocation policies toward a less interventionist state. At the *central* level, this phenomenon can be characterized by the structural change in government expenditures (see Table 6.1). Redistribution, indicated by the ratio of central subsidies and investments, is declining and consequently is restricting the administrative power of the government in the economy.

Defining the New Roles of Local Governments

The relationship of *local* finances and the economy is more complex. Direct economic influences and the regulatory role of local governments have diminished since the era of the Soviet-type council system in Hungary. But that does not mean a collapse of horizontal local fiscal and economic contacts. On the contrary, mutual dependence of local government and business is growing as state-owned property undergoes privatization and new local revenue structures are introduced.

The history of the laws regarding local governments illustrates the close connection between changes in municipal structure and the economy. After World War II, the first act concerning local governments in 1950 introduced the Soviet-type council system. The second act, in 1954, was a slow, weak step toward decentralization, indicating a power shift inside the Communist party stemming from economic problems of the command economy. The ideological basis of the third act (1971) originated in the economic reforms of 1968. In the 1980s, economic and political crises hastened the need for further reforms. In 1984, local administration was reorganized into a two-tier system; the "new" parliament approved the fundamental act on local governments in 1990 (Péteri, 1991b).

Proposals published during this last legislative process show the types of values involved in defining the economic role of local govern-

Table 6.1 Changes in the Structure of General Government Expenditures in Hungary, 1988-1991

Expenditure Category	*Changes in the % of GDP*
Economic services (subsidies, investments)	–10.1
Public services	3.9
Social security	2.0
Public debt	2.8
Others	–0.2
Total	–1.6

ments. In regard to municipal property, the crucial point of local fiscal autonomy, the first drafts declared municipalities to be the "owners of productive capital." According to this proposal, municipal property would include all the plots inside the administrative boundaries and the local companies. So municipalities could have been potential market actors. Later, another concept emerged in which municipalities were viewed as "public authorities" influencing the local economy by local taxation. This concept emphasized the need for market "neutrality," the danger of local monopolies, and the public service character of local governments. So a shift occurred away from the concept of municipal ownership to local governments as public authorities.

In fiscal terms, the act originally planned to transfer (in addition to the local revenues detailed below) part of profit taxes and value-added tax to local budgets. But given that these items would have increased differences among municipalities, this did not fit the concept of a "decentralized" local government system and was dropped. The assignment of municipal responsibilities was also a matter of debate. In the previous system, local functions were too widely defined; in the new act, municipal tasks needed to be separated from the central ones. In the first drafts, the separation of mandatory and facultative, optional functions and competencies was decided. In some versions of the act, the detailed separation of the mandatory and facultative functions came with an argument for municipal implementation but emphasized central provision of revenues to cover local mandates. This indicated an intent to separate local functions and to "deconcentrate" municipalities from the central state; subsequently, the purpose shifted from separation to a more general definition of tasks. The final version provided for a formal separation of mandatory and optional functions but without central budget grant responsibilities.

The New Fiscal Context

Local expenditures as a percentage of GDP show a rather wide range of local functions: 14.6% in the years 1986-1991. Mandatory tasks have the greatest share in local budgets. Education, health care, and social services together represent two-thirds of municipal expenditures (see Table 6.2).

Local functions define *revenue structures*. Large differences in local public service needs require higher shares of central grants; as general block grants are the major allocation method, this enhances local autonomy. The distribution of shared revenue has a double character: Personal income tax is a centrally levied, partly locally spent tax, transferred to the residence of the taxpayer, but per capita differences are equalized up to a minimum level. There is a trend toward a slowly declining ratio of central and shared revenues in local revenue structures, with increasing local fiscal and managerial discretion and pressures. As a consequence, municipalities are forced to raise their own revenues (see Table 6.3).

In this period of severe fiscal constraints in Hungary, local public finance regulations exhibit some of the peculiarities of centralized economies: There is an upper limit on total local expenditures, grants from the central budget are declining, and high municipal revenue obligations are in place (Council of Europe, 1986). Fiscal austerity has had a dampening effect on local capital expenditures and has challenged the efficiency of public service delivery. On the revenue side, simple (extensive) revenue-generation methods (collection of delayed taxes, transfers from corporations, short-term borrowing) and efforts to increase revenue capacity are also used (Péteri, 1989a).

The City-Market Context

Here is where local governments are linked to the market. Increasingly, economic conditions—fiscal regulations, local property and capital, rules of income distribution—define the success of local revenue-raising policies. The economic activity of municipalities is not only a simple for-profit entrepreneurial one. Local government also works as an administrative unit: It has influence through taxation, physical plan-

Table 6.2 Local Expenditures in Hungary, 1990 (in percentages)

Municipal Functions	*Current*	*Expenditures* *Capital*	*Total*
Health care	23	11	21
Social services	8	1	7
Education	42	17	38
Culture, sports	5	3	4
Public administration	6	3	5
Housing	2	20	3
City maintenance	4	7	4
Public transportation	7	10	7
Water, sewage	0	14	3
Other	3	14	6
Total expenditures	100	100	100
Current/capital expenditures	83	17	100

SOURCE: Ministry of Finance.

ning, and land-use policy. But in having effects on the local economy, municipalities are also faced with the problems of the economy. Thus there is a pressure coming from the market: When the local budget is connected to local resources, any changes (either depression or growth) will have strong effects on municipal service functions.

Regional Aspects of the Economic Crisis

The general crisis of the Hungarian economy can be characterized by the actual 5-7% decrease in GDP. This decline has a strong effect on municipalities because of declining final consumption, including public (current and capital) expenditures at the local level. There is high debt pressure on the economy, forcing economic policy to address the foreign trade balance. After the collapse of the Eastern European "markets," there is a simple claim not only for increasing exports, but for a rapid change in export orientation. Furthermore, during the past several years, there has been growing inflation, reaching a peak of 30% in 1991. Without adjustments in the inherited economic structure there will be

Table 6.3 Local Revenues in Hungary, 1991 (in billions HUF)

Revenues	*HUF*	*Percentage*
Own revenues	80.75	19.4
Local taxes	9.48	2.3
Shared personal income tax	47.02	11.3
Grants	190.68	45.7
General grants	155.46	37.3
Normative grants	148.53	35.6
Pit equalization	6.93	1.7
Specific grants	35.22	8.4
Specific grant	5.57	1.3
Addressed grant	11.77	2.8
Other	17.88	4.3
Social security fund	67.09	16.1
Credit, bonds	4.84	1.2
Other	26.54	6.4
Total	416.92	100.0

SOURCE: Ministry of Finance.

no changes in the output, but these adjustment processes contribute to rising unemployment.

In addition to these general economic problems, the *regional* effects of the crisis are crucial for local governments. In Hungary, industrial policy traditionally was concentrated and centralized. The *concentration,* or ratio, of large industrial plots is gradually declining, as the share of plots with 51-300 employees is increasing. Today the ratio of small companies (fewer than 100 employees) is relatively low (27.5%) relative to other European countries (above 45%) (*The Companies of the European Community,* 1990, cited by Kiss, 1992; on spatial structure of industry, see Table 6.4). This suggests that Hungarian municipalities must deal with larger companies, which makes cooperation and implementation of local economic development more difficult. The regional economic profile also shows strong *centralization*: 26% of the indus-

Table 6.4 Number of Plots and Employees by Size of Industrial Plots in Hungary, 1989

Number of Employees	*% of Industrial Plots*	*% of Employees*
1-50	61.4	7.0
51-300	28.3	37.1
301+	10.3	55.9
Total	100.0	100.0

SOURCE: *Statisztikai Evkönyv* (1989).

trial plots are located in the capital; corporations with headquarters in Budapest employ 34% of the industrial labor force, and one-third of their plots are in the countryside (1989).

Depressed areas are an additional regional economic feature. As identified by official regional policy, depressed areas include one-third of the towns and villages (97/1992 governmental decree). This high number of municipalities and their regional dispersion (in 7 of 19 counties) reveals the seriousness of the local problem, even though only 5% of the population lives in these communities. Industry is located in only one-fifth of these settlements, and one-fourth of the cooperatives in agriculture have deficits (*A gazdaságilag,* 1987). Besides the lack of local economic base, there is rising unemployment. At the beginning of 1991, the unemployment rate was only 2.7%; by the end of the year it increased to 9.5%. The numbers of unemployed relative to jobs are highest in counties where sectors of industry are in crisis or that traditionally have poor, agricultural economies (13%-18% unemployment in June 1992).

These regional economic variations suggest not only that conditions for municipalities differ, but that, for many of them, there is no real choice in local development and fiscal policy. At this point, the centralized structure of industrial plots and unemployment are interconnected: Companies close or sell their factories and plots first in the countryside. There is a high rate of daily commuting in Hungary, so workers from remote villages tend to lose their jobs first. The structure of unemployment coincides with differences in urbanization (with the "urban slope"): 83% of the unemployed are physical (blue-collar) workers, 54% are unskilled, and these numbers are similar to the structure of the labor force in villages (*Munkaerôpiaci Helyzetkép,* 1991).

The Privatization Response

The most uncertain element of economic transition is privatization. Dismantling of state-owned property began in the 1980s, when a new regulation for establishing market-type corporations, partly by transforming state companies, was approved. Up until 1990, when centrally directed privatization began, the number of small companies was increasing. As organizational forms of firms are limited, small enterprises were the highest numerically, whereas trusts and big state firms remained almost unchanged. The earlier tax exemptions given to foreign capital caused a rather high share of such investments (43%-47%). In 1990, joint ventures numbered 5,571, with 3,814 new ventures created in a single year; the average value of invested capital in these corporations, however, is rather low (below 10 million HUF) (Csillag, 1991). Similar to the high number of joint ventures, according to a State Property Agency publication, by May 1992 two-thirds of the 5,000 shops were also privatized (*Privinfo,* vol. 1, no. 6, 1992).

Other privatization programs are going on with various degrees of involvement by the State Property Agency. Some privatized companies are "transformed" to corporations without any major foreign or domestic investment. The state-initiated "active" privatizations resulted in 45 transformations; self-privatization (without state approval) yielded 85 units. Company-initiated privatization of entire or partial units has transformed 350 billion HUF. Despite the fact that in 1991 foreign investments in Hungary exceeded $1.8 billion—the highest in the region—the privatization process has slowed down. The two contradictory goals of the State Property Agency (increasing budget revenues through privatization and avoiding giveaways) make the agency more influential in the privatization process.

Regional aspects of privatization, as an innovative process, are important in estimating the future of local economic development policy. Local governments have a 2% share of privatization revenues; privatization follows existing differences in urbanization. Some forms of new, small companies are centralized in the capital: One-third to half of them were established there, and 37% of the new value is produced there. Wages and salaries are higher there than the average (new value per capita, 133% of the average; new value per capita, 131%). In villages and in depressed areas, entrepreneurial activity can hardly be found (Juhász & Kamarás, 1990). Other research has shown that diffusion of innovations is determined not only by "distance factors" (Ruttkay & Nemes-Nagy, 1989). Regional differences in innovation are explained best by the size hierarchy and administrative status of the settlements. Regional differences show a diffusion of "innovations" starting in the western (Transdanubian) regions and moving toward the eastern ones (Great Plain, Nordic Hills).

So the regional consequences of the economic crisis—assuming closer economic and financial contacts—affect municipalities and cannot be separated from economic problems of regional differentiation. In this respect, conditions of resource allocation, formed by history and externalities, strongly determine municipal finances. This pressure comes to the local level from outside, although some elements of it can be influenced by municipalities. The subsequent question is, How effectively can municipalities change their economic and infrastructural conditions? Are there sufficient *possibilities* for local governments to modify the allocation trend and promote local economic development?

The Emerging Relations of Municipalities and the Market

Local governments are faced with for-profit activities in three different ways: as owners (entrepreneurs), as public authorities, and as administrative units providing services. The consequences of these three options for the local economy are different, and the policy choices depend on local circumstances.

As the struggle over the definition of municipal property revealed, local governments, contrary to the original intentions, do not possess significant productive capital. Nor are they the winners in the privatization process, because central budget officials are not willing to give privatization revenues to municipalities. Even the gains from privatizing local public enterprises are only partially directed to municipalities. After detailed rules of property transfer were set up by the Parliament, it became clear that municipalities possess mainly infrastructural assets. Formerly state-owned public (social) housing and other buildings (stores, small shops, garages), vacant areas, historical buildings, water flows, and national parks previously managed by the council became "municipalized." The process of property transfer shows that municipalities will not have a direct effect on the local economy through control of productive assets.

Nevertheless, the majority of municipalities are starting active local *entrepreneurial* policies. Mayors utilized this managerial autonomy particularly during the early years of fiscal restrictions in order to gain greater maneuverability. As economic circumstances changed, entrepreneurial activity developed; municipalities with property and buildings got in touch with investors. In the late 1980s fiscal policy expressly supported "entrepreneurial municipalities" because of the need for additional local source revenues. The number and economic weight of these corporations, in terms of municipal capital investment, however, were not significant. Closer to the political changes in 1988-1989, local leaders were accused of transferring their political power for economic gain. Indeed, in many cases, as private persons they had shares in these municipal companies or worked as managers of them, giving up their political careers. Local acceptance of these activities varied.

Local governments also are connected to the economy through their *revenue policies*. As shown in Table 6.3, the proportion of own revenues in local budgets is rather low, but it can be important for a municipality. This group of revenue items is in transition from being centrally prescribed, locally collected fees and user charges to real local revenues and taxes. The importance of own source revenues is real: In 1991 actual revenues were 23% higher than the planned ones. In this respect, local taxation holds great possibilities. Local taxation serves as a general fee for municipal services and local infrastructure. Local tax, measuring externalities, is a factor cost, so it has an effect on resource allocation. Obviously, it has an influence not only on production but on income distribution as well, by transferring income in a municipality via the local budget.

As a consequence of the Hungarian concept of taxation, local tax sources include property, tourism, labor, and business. These distinct potential local tax revenues follow the logic of separation of local and national functions. The system allows for strict controls at the macro level because local taxation sources and maximum rates are centrally defined but give municipal discretion, too. The act allows for different categories of local taxes and for municipal choice among six different local taxes, within maximum limits; methods of assessment are also optional, but defined by the act. The local government decides on exemptions in terms of income, parts of the city, social status, or any other criteria. These are municipal taxes, because county (regional) local governments have no right to levy taxes. The municipalities are responsible for collection, so the central organization established for state taxation in 1988 gives information but has no duties in valuation, assessment, and collection. Local taxes are deductible from the tax base of personal income tax; for companies and entrepreneurs, they constitute a cost item. For political and technical reasons, municipalities levy local taxes primarily on business and tourism, because the collection of these taxes is easier; they do not burden personal income directly.[3] In fact, in 1991 only one-tenth of municipalities levied local taxes (Péteri, 1991c).

Previously (1981-1987), local governments were also on the market as *bond issuers* (Lados, 1991). Municipalities issued bonds to finance their own infrastructural investments when personal savings and the rate of inflation were rather stable and low. Local bond issues were mainly in the field of public services (education, health care, public utilities), to supplement central redistribution; they were used more frequently in municipalities where the grants were relatively low (Péteri, 1989b).

The long-term benefits of *administrative measures* are still not used by the municipalities. Local physical (master) planning is separated from other municipal functions and driven by other professional values. Because of the short-term fiscal needs, land-use planning is not integrated into local policy.

Finally, local governments as *public service units* are also linked to the local economy through infrastructural development policy and also because of the collapse of public institutions responsible for earlier forms of service provision. The traditional concept of a centralized public service delivery system allowed only directly controlled organizations, operating without any economic motives. As the budget revenues declined, a slow "privatization" of public services began. The number of nonprofit institutions is gradually increasing, and now they are subsidized by the central budget. Other forms of service provision (e.g., contracting out in city maintenance) also have direct influence on the local economy.

These market links of local governments have various positive effects on the economy. As owner of capital and as entrepreneur, the municipality is able to influence the local economy directly. This direct effect is weak, but as state property becomes subject to privatization, municipal possessions can be important. Local revenue policy—by fiscal and administrative measures—has an allocative function in helping set a more effective investment policy for the entire economy. Methods of local spending also influence the market through new forms of municipal public service delivery.

Political and Economic Implications

Expanding local relations in an emerging market economy coincide with economic crisis and thus strengthen the contradictory effects of the economy on local governments. For some municipalities this market pressure harms their income distribution status and potential, whereas for others with positive externalities, these changes could be profitable.

On the other hand, an active local development policy causes more efficient choices in investment policy for the entire economy. When a municipality influences economic conditions by correcting allocative decisions, it raises the efficiency of productive means. Local governments, as simultaneously economic *and* public service (political) organizations, are in a position to compel long-term social, environmental, and more abstract values, even though immediate pressures for revenue-raising policy may determine local economic contacts.

The possibilities of local governments are expanding through fiscal decentralization and "privatization" (municipalization). Case studies describing the transitional period for municipalities illustrate a polarization of local fiscal policies (Péteri, 1991a);[4] that is, viewing municipal fiscal strain as an adaptive ability (Clark, 1986), two basic values in response to market challenges appear to determine local fiscal policies, including economic development policy, in Hungary (Péteri, 1992). Municipalities supporting the "public manager/public authority" concept have stronger expectations about central grants; mayors in these municipalities have hostile attitudes toward privatization, and profit-making property is not important to them. Local officials expressing these views are found basically in municipalities located in the Transdanubian region of the country (where the Somogy case is located). This part of the country is in a better situation than others; there local governments are not forced to be involved in economic development. In contrast, municipalities acting as direct economic actors are more frequently found in regions hit by the crisis (Hungarian Plain, Nordic Hills). Mayors there support municipal economic functions (development, employment), claim local property, and intend to participate in local economic activities as entrepreneurs.

Local fiscal policy is a social phenomenon, as is political culture (Clark & Inglehart, 1990), so there are close relationships among the objectives of local economic development policy, the degree of municipal activity, and local public service methods. As local governments face the market, new functional objects come to the fore, and leaders and administrators learn a new fiscal mentality. In Hungary, this feedback mechanism has very important effects in the new decentralized local government system. It is possible to build a direct link between the two main municipal activities (public service delivery and revenue-raising activities). In a market economy this is not an administrative relation, but one based on a more equal partnership. Municipalities strengthened by political and economic means are not subordinated to other economic actors, so they are able to have positive effects on allocation.

Notes

1. In local finances two special types of revenues demonstrate the idea of self-financing: direct transfers of companies to local budgets (for maintaining nurseries and kindergartens and for housing) and special wage taxes sent to the local budgets. The latter diminished in 1990, the direct transfers declining as companies gave up these types of subsidies because of economic crises.

2. Direct contacts of the economy and public sector occur through taxation. In Hungary, the personal income tax and value-added tax have existed only since 1988, so up to that time the corporate (profit) tax was the typical form. This close relationship between direct corporate taxes and public expenditures strengthened and at the same time verified the dependence of the economy and administration, and so the need for direct intervention. Nowadays the taxation structure has changed, but corporate tax and social security contributions paid by employers are high (6.1% and 17.2% of GDP); the rate of personal income tax is low (7.1% of GDP).

3. The personal income tax cannot be regarded as a local tax, although in 1991 one-half of personal income tax revenues were transferred to local budgets. The tax base, rates, exemptions, and percentage of sharing are centrally decided. There is an equalizing mechanism connected to differences in specific personal income tax, but there is an indirect and lagged relationship between the local economy and municipal budgets through changes in personal income levels.

4. In the county of Somogy, a closer and more active connection between voters and municipalities is attempted. In this relatively rich county, where the strong county local governments have built up a proportional settlement structure, municipalities could raise high levels of own source revenues. Here, party programs laid a claim for a "managerial" municipality, *providing public services.* Based on this idea, local leaders and employees should map local needs and increase the efficiency of the municipality (a "busy, industrious representative body"). They have practical, professional programs fulfilling their electoral duties, but also have the right to manage local affairs.

In contrast, in Békés, the previous local government built up a decentralized fiscal redistributive system that laid stress on cities and towns. In this region of low grants there are many municipalities with fiscal strain. Searching for local revenue sources, municipalities count on private incomes. The formation of new local parties and municipal leaders emphasized expectations for close contacts with the voters, building on direct participation. Local government, first of all, is not only a unit of public service delivery, but also *provides benefits* in other areas of private life. It sought to build conditions for entrepreneurs, to survive on the municipalities' own assets, and to privatize or contract out its functions.

References

Clark, T. N. (1986). *Municipal fiscal strain.* Chicago: University of Chicago.

Clark, T. N., & Inglehart, R. (1990). *The new political culture.* Paper presented at the biennial meeting of the International Sociological Association, Madrid.

Council of Europe. (1986). *The response of local authorities to central government incitement to reduce expenditures* (Study Series 37). Strasbourg, France: Author.

Csillag, I. (1991). Magyarország [Hungary]. In *Privatizáció Kelet-Európában* [Privatization in Eastern Europe]. Budapest: Atlantisz-Medvetánc.

A gazdaságilag elmaradott térségek társadalmi-gazdasági jellemzöi [Economic and social characteristics of depressed areas]. (1987). Budapest: Központi Statisztikai Hivatal.

Juhász, I., & Kamarás, G. (1990). *Piaci vállalkozások területi jellemzôi* [Regional characteristics of market ventures]. APEH.

Kiss, S. J. (1992, August). Kis és középvállalkozások elindultak és . . . ? [Small- and medium-sized enterprises started and . . . ?] *Magyar Hirlap.*

Kornai, J. (1982). *A hiány* [The shortage]. Budapest: KJK.

Lados, M. (1991). *A tanácsi kötvénykibocsátások tapsztalatai Magyarországon (1981-1988)* [Experiences of municipal bond issues in Hungary]. MTA RKK Györ.

Munkaerôpiaci Helyzetkép [Report on the labor market]. (1991). OMK No. 2.

Péteri, G. (1989a). A költségvetési restrikció hatása a tanácsi gazdaságra [The effects of fiscal restriction on local budgets]. *Ållam és Igazgatás, 10.*

Péteri, G. (1989b). A tanácsi gazdálkodás És a kereskedelmi bankok kapcsolata [Connection of local finances and the commercial banks]. *Bankszemle, 4.*

Péteri, G. (1991a). Budget in the background. In *Events and changes: Local transition in East Central Europe* (LDI Project Working Papers).

Péteri, G. (1991b). *Changes of concepts: Legislation on local governments, 1987-1990.* Paper presented at a workshop of the International Political Science Association, Budapest.

Péteri, G. (1991c). *Local taxation in Hungary.* Paper presented at the IRRV European Conference, Paris.

Péteri, G. (1992). *Responses to local challenges.* Paper presented at the annual meeting of the Midwest Political Science Association, Chicago.

Ruttkay, É., & Nemes-Nagy, J. (1989). *A második gazdaság földrajza* [Geography of the second economy]. OT TGI.

Statisztikai Evkönyv [Statistical yearbook]. (1989). Budapest: Központi Statisztikai Hivatal.

Vági, G. (1991). *Magunk, uraim* [We, ourselves]. Budapest: Gondolat.

Wiener, G. (1982). A gazdasági körzetek [Economic regions]. *Területi Kutatások, 5.*

7

Regimes of Accumulation, the Caribbean Basin Initiative, and Export Processing Zones

Scales of Influence on Caribbean Development

THOMAS KLAK
JAMIE RULLI

The pace of political economic change is quickening worldwide. Recent events in contexts as diverse as Russia, Haiti, and California reveal that regional political economic fortunes can deteriorate in only months. Investment flows can converge equally quickly upon a region, as the flood of U.S. corporate prospectors to Mexico searching for profitable opportunities in anticipation of NAFTA illustrates.[1] And, as large firms increasingly function as international rather than national actors (Dicken, 1992, pp. 3-4), governments scramble to join regional trading alliances with hopes of avoiding exclusion from transnational business and trade.

These and other political economic transformations and realignments across global regions are proceeding faster than our ability to understand the contributions that processes operating at the various geographic scales make to the local outcomes. On one hand, we know that global processes, such as recession and hypercompetition, shape conditions in all corners of the globe. On the other hand, at certain historical junctures *the local can become the global,* as the spread of Japanese manufacturing systems from one rather isolated locale to factory floors worldwide demonstrates (Storper & Walker, 1989). Between these extremes are myriad less distinguished local policies, international tariff agreements, and economic trends contributing to the broadly defined development prospects of lesser-known regions. To exercise some local control in such a turbulent global context, many are discussing the potential of a *new localism,* which has become a core concept of those interested in *empowering* localities and local governments as well as in *measuring* their power against economic and larger-scale forces (Logan & Swanstrom, 1990; Mayer, 1989). Either version of the new localism requires that the contributions of influences shaping the political economic fate of regions be delineated and understood. This chapter offers such an effort for a sampling of Caribbean countries, which have increasingly oriented their economies toward the role of industrial export platforms for U.S. investors.

The substantial literature on recent Caribbean industrialization can be organized around three types of explanation, each representing factors operating at a different scale in the global political economic system. The first interpretation is that growth in Caribbean industrial exports results from macroeconomic forces or, more specifically, profit cycles and the evolution of the dominant regime of accumulation. A second scale of explanatory factors is at the level of policy from developed countries; in the case of the Caribbean, the U.S. Caribbean Basin Initiative (CBI) is most crucial in this regard, and refers to several layers of U.S. trade policy toward the region. Yet a third scale of influence can be identified in the policies of Caribbean countries themselves, and represents the potential contribution of a new localism to economic restructuring. Whereas political elites *articulate* local policy promoting a country as an industrial export platform, policy embodies the pressures and interests of others in and out of the government and the country while it neglects the interests of others (Offe, 1975). The social struggle underlying such policies requires detailed empirical investigation (Hey, in press), particularly in the Caribbean, where it has scarcely been examined. Until then, we must judge the representativeness of local policy in terms of who benefits from the outcomes.

Within the Caribbean Basin, we examine the industrial exports of eight countries: Barbados, Dominica, the Dominican Republic, Grenada, Haiti, Jamaica, St. Lucia, and Trinidad and Tobago. These countries have hosted the vast majority of recent industrial development in the region.[2] Despite its membership in Reagan's Caribbean Basin designation, we exclude the war-torn and analytically distinct mainland portion of the region from this empirical analysis (Newfarmer, 1986). Our eight nations represent the range of Caribbean countries in terms of culture, race, language, size, wealth, and industrial development (West & Augelli, 1989); contain most of the region's population in countries trading with the United States; and all are following development paths emphasizing economic ties with, and exports to, the United States.

The Context of Recent Caribbean Industrial Development

As our delineation of three scales of explanatory factors suggests, the trend of U.S. firms relocating production to Caribbean islands results from several intersecting forces, about which we would like now to be more specific. Intense global competition has encouraged firms in developed countries to cut production costs by moving labor-intensive aspects of production to relatively low-wage regions such as the Caribbean (Codrington & Worrell, 1989; Schoepfle & Perez-Lopez, 1989; Wilson, 1990). Policy makers in both developed and Caribbean countries, influenced by the success of offshore assembly in Asian newly industrializing countries (NICs), generally agree that industrial exports through multinational corporation (MNC) investment is a worthwhile development strategy for the Caribbean (Deere et al., 1990, p. 40; Griffith, 1987, p. 61). This strategy is also sanctioned by major international financial agencies, such as the International Monetary Fund (IMF) and World Bank (Codrington & Worrell, 1989; Dicken, 1992). From the viewpoint of powerful interests in developed or peripheral countries that favor the social status quo, industrial product export has the distinct advantage that grave social inequalities need not be addressed for there to be economic growth (Thomas, 1988). Thus Caribbean export industrialization is fueled by an alignment of interests among political and economic powers from both core and periphery.

We focus on the Caribbean for several reasons. The islands share with their Latin American mainland neighbors historical economic dependence on core countries and contemporary foreign debt (Golub, 1991). Other economic problems, however, make the development prospects of the Caribbean more elusive than for most of Latin America. The small internal markets and grave inequalities in the island countries of the

Caribbean, in this sense similar to their South Pacific island counterparts, have dampened gains from MNC-led import substitution industrialization, or ISI (Codrington & Worrell, 1989, p. 29; Griffith, 1990, pp. 847-849; McKee & Tisdell, 1990, p. 10). The limitations of ISI (see Kay, 1989) and the lack of capital in small island countries have pushed them toward the front of the pack of countries aggressively adopting development policies featuring MNC-led industrial assembly for export (McKee & Tisdell, 1990, pp. 18-19). A Caribbean focus is also useful, considering that the Americas from Canada to Chile are increasingly intertwined in terms of industrial production (Dicken, 1992, pp. 171-172). Within the American hemisphere, Caribbean islands have surplus labor, low wages, and favorable accessibility to U.S. markets. Observing these conditions, the Reagan administration in the early 1980s designated "the Caribbean Basin" a new trade region that includes the Caribbean islands, Central America, and the Guyanas.[3] In combination, these factors offer the Caribbean Basin great potential for new industrial exports. At the same time, they make the region a particularly important empirical context for understanding the importance of local policy in contemporary economic processes that increasingly intertwine First and Third World nations.

Other economic weaknesses in the Caribbean also encourage policies that promote industrial exports. Since the 1970s, Caribbean countries have experienced falling earnings from their traditional export commodities and rising interest owed on foreign debt ("Development," 1992). To earn foreign exchange, service their debts, create jobs, and diversify their economies, they have engaged in economic development activism in the form of new export policies, particularly for industry (Hillcoat & Quenan, 1991, p. 194; Schoepfle & Perez-Lopez, 1989, p. 132). This type of industrialization primarily features establishing branch plants or subsidiaries of developed-country firms that conduct their labor-intensive production processes in the Caribbean (Codrington & Worrell, 1989, p. 31). Because most of these operations consist of product assembly, we have created the acronym PACE, for "product assembly in the Caribbean for export," to refer to industry in the islands.

To encourage PACE, the U.S. government has crafted region-specific policies, such as lower import tariffs on certain Caribbean industrial products. Canada and the European Community have similar policies toward the Caribbean, but the combination of U.S. economic power, its proximity to the islands, and its authority over the Caribbean during the twentieth century leads to the United States having by far the most investments in, exports to, and imports from the region (Deere et al., 1990, pp. 162-164). The economic importance of the United States for the Caribbean is not diminished by the greater share of investment and trade between industrialized countries than First and Third World countries (Gordon, 1988). Thus an important regional-industrial bloc joins the United States and the Caribbean, with the latter playing a growing role as provider of assembled products that are controlled and consumed in the former. Similarly structured regional-industrial blocs are more closely tying the Japanese economy to those of Southwest Asian countries, and Western European economies to those of Eastern Europe. A major contribution of Caribbean governments to the promotion of PACE has been their creation of special extraterritorial industrial districts called *export processing zones* (EPZs), which offer foreign investors tax incentives and other financial benefits (Dicken, 1992, pp. 181-183).

Given the huge disparities of power and influence separating the policy-making apparatuses of a core country, such as the United States, from its peripheral trading partners on tiny Caribbean islands, the concern in this volume about the changing *vertical linkage* between levels of government takes a different twist. We, too, are concerned with the distribution of authority over policy and economic outcomes between political entities, but ours are mismatched nation-states rather than central and local governments. The populations of Caribbean countries range from 40,000 for St. Kitts-Nevis to 10.8 million for Cuba, and therefore these countries approximate the size, and perhaps the political power, of local government units in the United States.

Policy initiatives promoting industrial exports are increasingly common across the Caribbean, and the region is clearly diversifying its industrial contributions to the global economy. In this chapter we investigate the extent to which these industrial policy and export trends are the result of a new localism in the Caribbean. In other words, in the face of rapid global economic change, most of which does not bode well for Third World countries (Storper, 1991), do the policies of Caribbean governments for industrial exports signify a notable move toward controlling their economic fates? Or are Caribbean countries essentially responding to pressures and interests outside their borders?

During the 1980s, PACE, when measured by its contribution to growth in jobs or in the value of exports, was the most dynamic sector of Caribbean economies (Deere et al., 1990, p. 99; Schoepfle & Perez-Lopez, 1989). From 1983 to 1987, U.S. imports of assembled products from CBI countries grew by 10%, compared with a 9% decline for all

CBI imports, led by falling prices and shrinking quotas for traditional primary products (Schoepfle & Perez-Lopez, 1989, p. 142). In only four years, assembled products' share of total Caribbean exports grew from 6.3% in 1983 to 13% in 1986 (Schoepfle & Perez-Lopez, 1989, p. 148). By 1988, U.S. imports of Caribbean industrial products exceeded $1.2 billion, a level two and one-half times higher than just five years previous (Deere et al., 1990, p. 158). Given this dynamism, we focus on PACE rather than on other important Caribbean economic sectors, such as mining (primarily petroleum and bauxite), agriculture (primarily sugar and bananas), tourism, financial services, and industry for domestic consumption. Trade and investment-stimulating policies on the part of the U.S. or Caribbean governments also have effects on other economic sectors, although arguably PACE has been the primary target of policy and has been most affected by it.

Recent Caribbean Industrial Export Trends

To investigate the forces underlying PACE, we define *industry* as whole or partial product assembly, and *assembly* as those industrial processes involving significant transformation of materials, although not necessarily into finished consumer goods. Our main data source (United Nations, 1980, 1984, 1989, 1990) reports annual exports per commodity for each Caribbean country, and separates products into 10 groups called Standard International Trade Classifications (SITCs). Our definitions of *industry* and *assembly* are captured in SITC Categories 5 (Chemicals), 6 (Basic Manufactures), 7 (Machines, Transport Equipment), and 8 (Miscellaneous Manufactured Goods). These four categories contain the bulk of manufactured exports and exclude agricultural and other primary-product exports. For the Caribbean countries examined in this chapter, Category 7 is mainly electronics and Category 8 is primarily apparel. We examine industrial exports for all years from 1975 to the most recent year for which data are available.[4]

Much valid criticism has been leveled against research on industrial processes in Latin America and the Caribbean for measuring it by way of official statistics on the size of economic sectors or the formal workforce. These statistics neglect the informal economic activities that feed subcontracted products into formal firms and support large portions of national workforces. Because we evaluate industrialization by examining the total value of exports, our data improve on most of the literature by capturing the informal contributions to production that are embedded in international industrial flows (Lawson, 1992; Safa, 1986; Thomas, 1988, pp. 93-96).

Another problem with most of the earlier Caribbean industrial studies follows from the fact that Caribbean countries range widely in population size (Table 7.1) and in the contributions of natural resource exploitation to GDP and employment, as illustrated by Trinidad and Tobago's regionally unique petroleum exports. Given this variation, measuring industry in the usual way, by its share of GDP, the absolute number of workers, or the absolute value of exports, does not effectively convey industry's local impact (Thomas, 1988, pp. 88-89). To standardize for the variation in the sizes and populations of Caribbean countries, we divided the total export earnings for each category by the country's population for 1980-1982 (United Nations, 1987). This per capita presentation of the data gives it uniformity and a meaningful scale. To compare Caribbean economies to a mature, diverse export economy, we included a comparable plot of U.S. export performance for the same four industrial categories (Figures 7.1-7.10).

The per capita industrial exports of all eight Caribbean countries (Figure 7.1), which is an annual sum of the exports of countries for which data are available per year, show a fairly regular pattern of growth from the mid-1970s to the end of the 1980s. At the level of individual Caribbean countries, however, the industrial export trends indicate much less consistent growth over time and in comparison to other countries (Figures 7.2-7.9). The main regularity among the countries is that they generally exhibit peaks and troughs in export performance. Even that pattern varies, however, because different countries experience growth and decline in different years (Table 7.1). The short-term peakedness in the time series for industrial exports suggests that investment irregularly flows in and is rapidly withdrawn from Caribbean countries. A brief review of the export products and trends for each of the eight Caribbean countries will conclude with our establishing an empirical base from which to examine possible explanations.

We define *export instability* as a significant change in total per capita export earnings, a situation that applies to most countries of the Caribbean (Table 7.1). Most Caribbean export instability results from the fact that individual countries depend primarily on a single industrial sector. In Barbados, for example, per capita export revenues were most severely affected in the rise and decline of its electronics industry, 1980-1987 (Figure 7.2). In the Dominican Republic, instability came primarily from

(Text continued on page 130)

Table 7.1 Major Trends in Recent Caribbean Industrial Exports

Country	*Population, 1980-1982*	*Per Capita GDP 1990 U.S.$*	*Largest Category*	*Peak Year*	*Trough Year*	*Range*[a] *(U.S. $1,000s)*
Barbados	252,029	6,540	electronics	1984	1987	151,573
Dominica	74,625	1,940	chemicals	1982	1984	2,822
Dominican Republic	5,647,977	820	manufacturing	1979	1982	94,073
Grenada	89,088	2,120	apparel	1982	1987	1,858
Haiti	5,053,792	370	apparel	1981	1975	40,483
Jamaica	2,205,507	1,510	chemicals	1980	1981	536,318
St. Lucia	115,153	1,900	electronics	1980	1985	8,356
Trinidad and Tobago	1,079,791	3,470	chemicals	1985	1987	70,779

SOURCE: United Nations (1980, 1984, 1989, 1990); Population Reference Bureau (1992).
a. The difference between the peak and trough years for the export value of the country's largest industrial category.

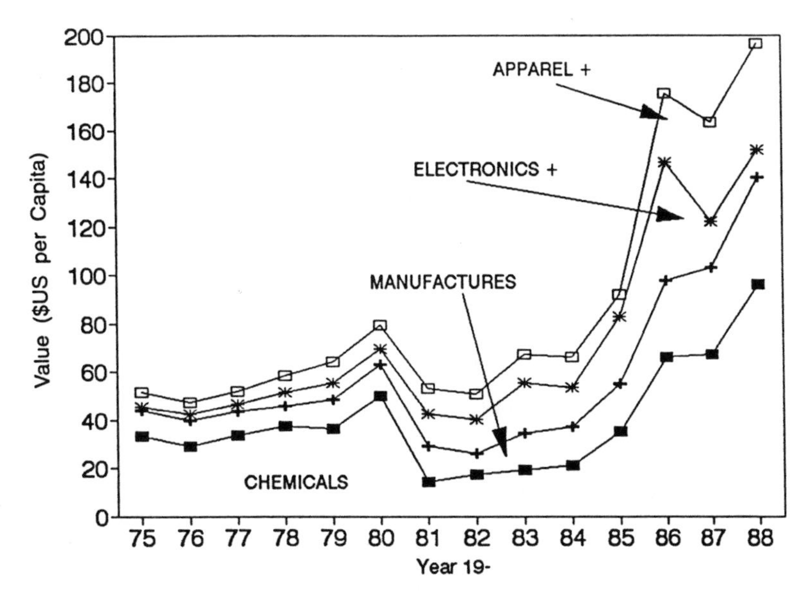

Figure 7.1. Per Capita Industrial Exports for All Eight Countries Combined

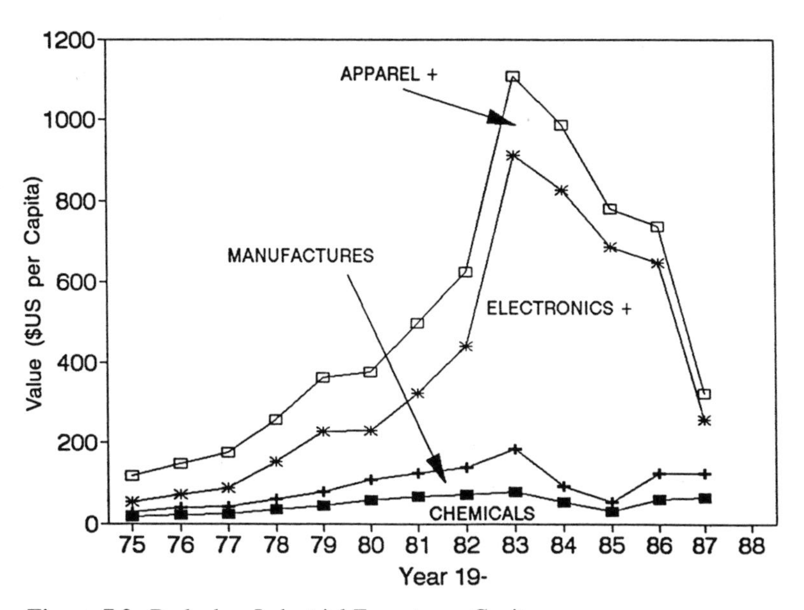

Figure 7.2. Barbados: Industrial Exports per Capita

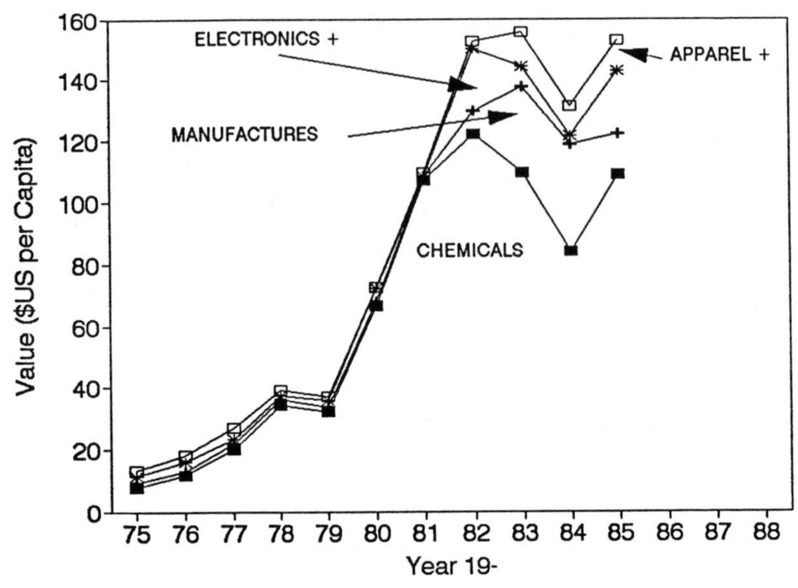

Figure 7.3. Dominica: Industrial Exports per Capita

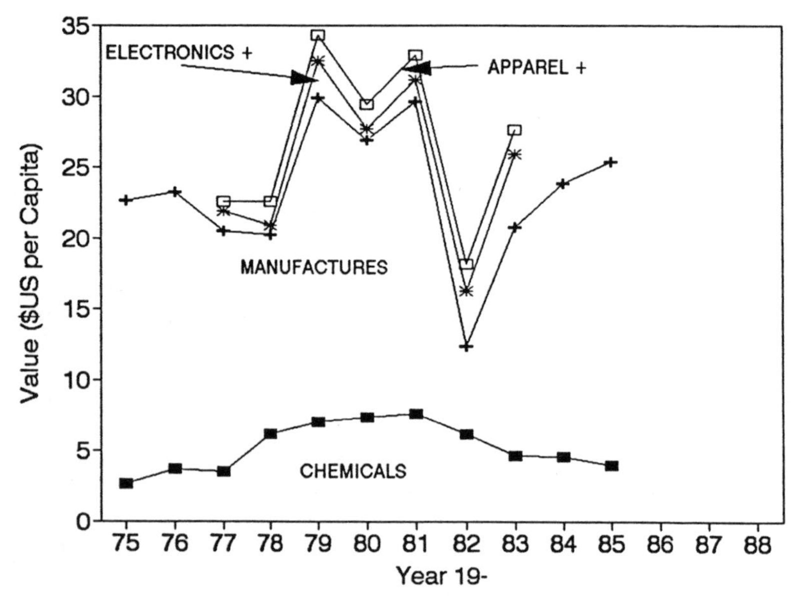

Figure 7.4. Dominican Republic: Industrial Exports per Capita

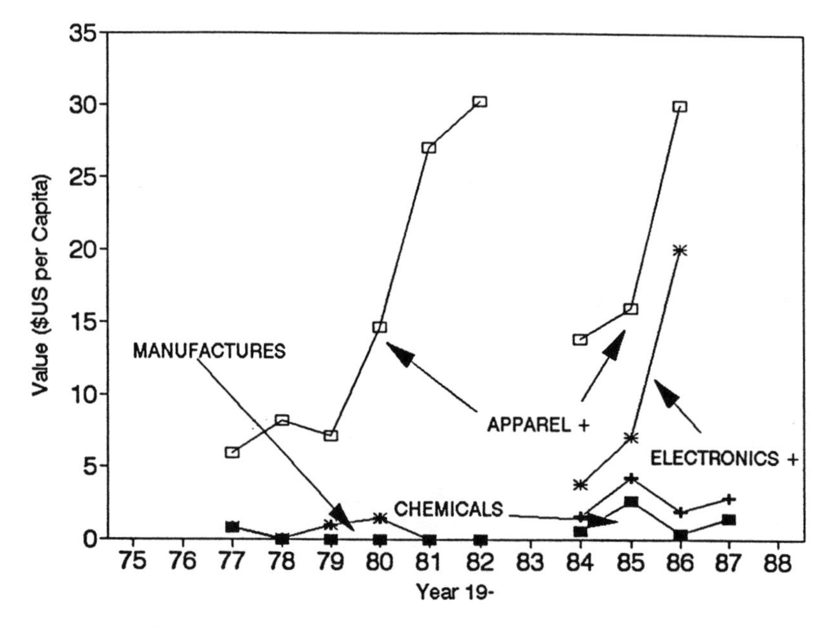

Figure 7.5. Grenada: Industrial Exports per Capita

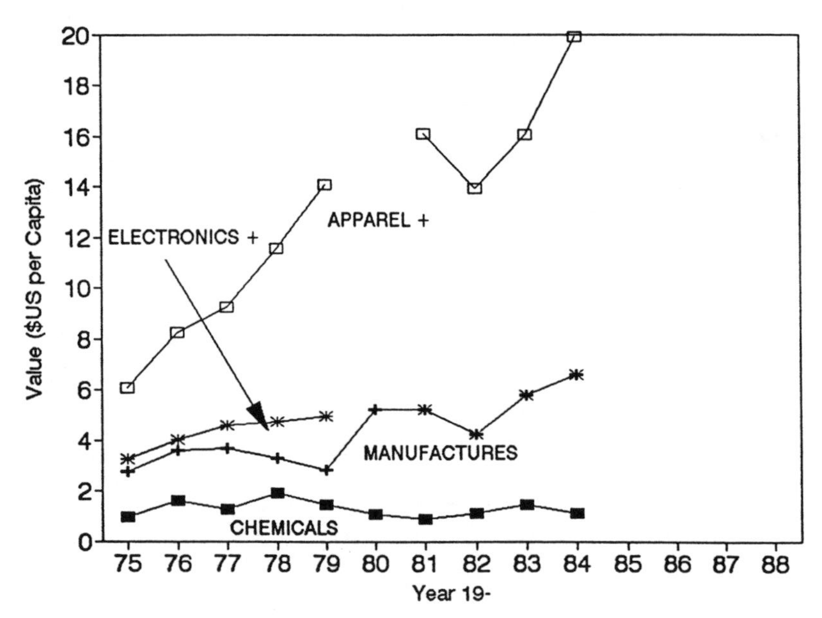

Figure 7.6. Haiti: Industrial Exports per Capita

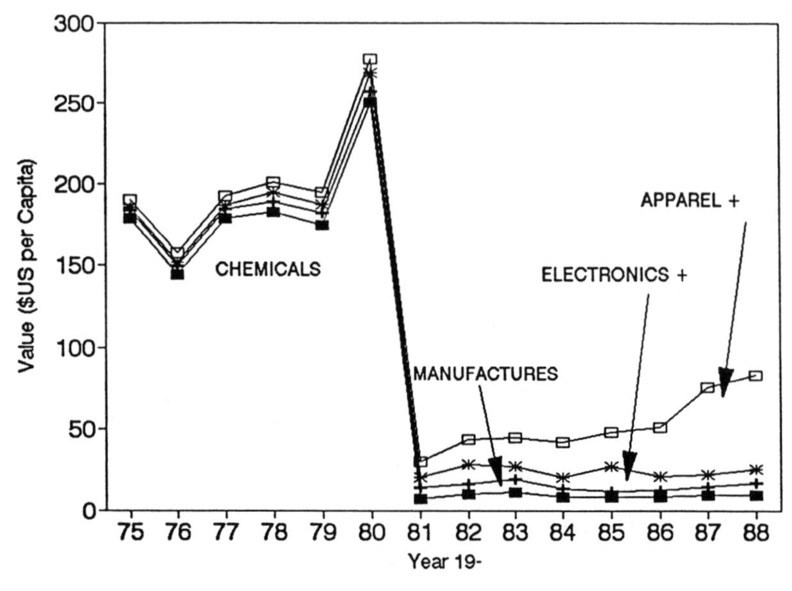

Figure 7.7. Jamaica: Industrial Exports per Capita

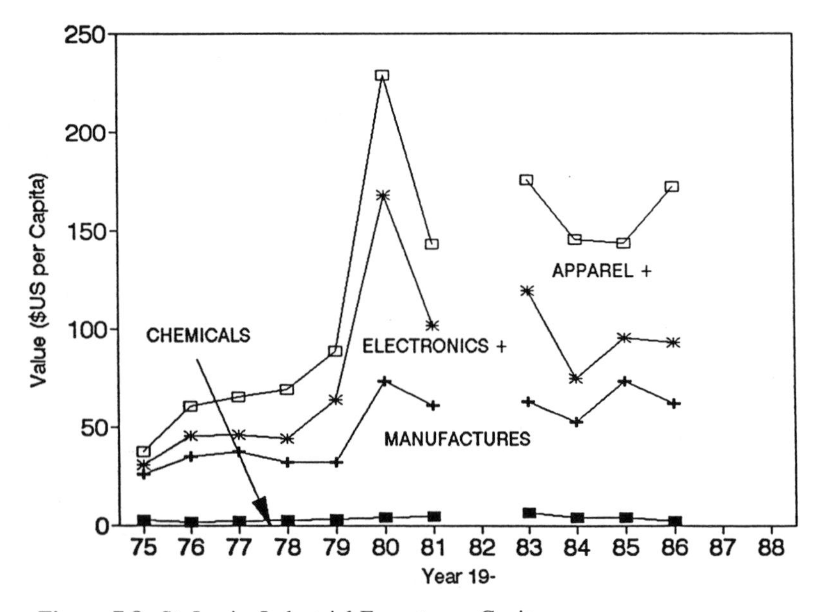

Figure 7.8. St. Lucia: Industrial Exports per Capita

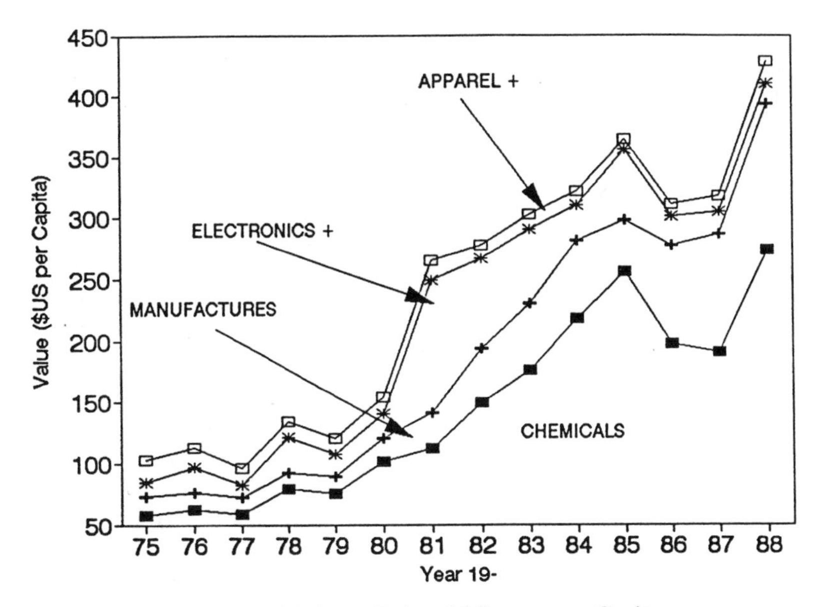

Figure 7.9. Trinidad and Tobago: Industrial Exports per Capita

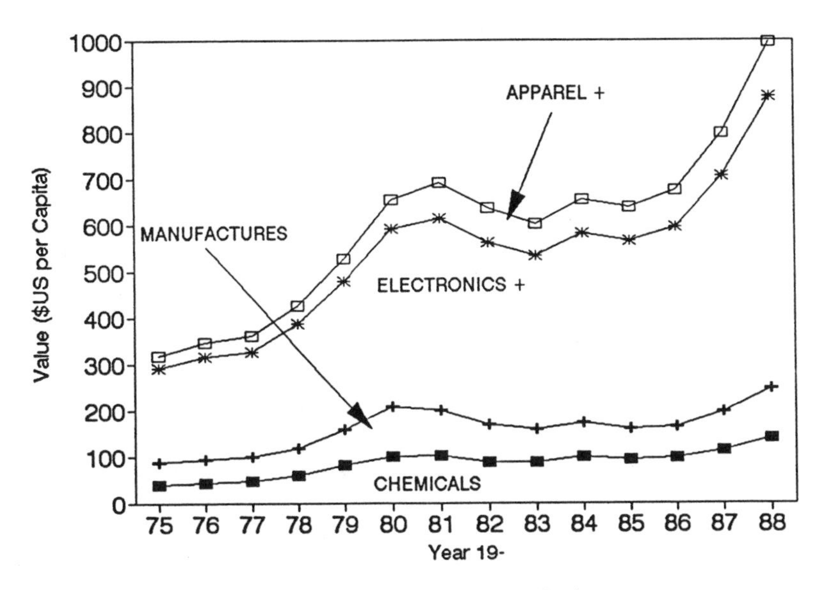

Figure 7.10. United States: Industrial Exports per Capita

a drop in iron-product exports from 1979 to 1982, which then rebounded, so that by 1985 they approached their 1979 value (Figure 7.4). In Grenada, apparel exports rose from $451,000 in 1977 to more than $2.6 million in 1982, despite provocations from the United States leading up to the 1983 invasion. By 1987, however, apparel exports had declined to $729,000 (Figure 7.5). In Jamaica, total industrial exports were unstable because aluminum exports, which were primarily aluminum oxides and hydroxides, plunged from more than $550 million in 1980 to just over $16 million a year later (Figure 7.7). Dominica's and St. Lucia's electronics exports also underwent severe fluctuations (Figures 7.3 and 7.8). In Dominica, the value of these exports jumped from only $1,000 in 1981 to more than $1.5 million the next year. By 1984, however, Dominica's total exports in this category were down to $198,000. St. Lucia's total electronics exports climbed from $977,000 to nearly $11 million in just three years, from 1977 to 1980. This increase was largely caused by the export of excavation equipment. However, by 1986, St. Lucia's electronics exports plunged to about $3.6 million.

In our sample, only a few countries experienced sustained industrial export growth in any sector. Dominica has steadily increased its exports of soaps, which are part of the chemicals category (Figure 7.3). Total soap exports grew from about $2.4 million in 1979 to more than $9 million in 1982, a significant gain in per capita terms. Soap exports dropped after that peak year, but remained the largest contributor to Dominica's exports. Haiti achieved sustained growth in the broadly defined apparel category as a result of sporting goods exports.[5] Despite this growth and Haiti's reputation as the major source of baseballs for the United States, however, the country's industrial exports are small in per capita terms, indicating that only a tiny portion of the population works in and benefits from export industry.

Thus Haiti's industrial exports grew but were still minor in per capita terms. All of the other countries except for Trinidad and Tobago experienced sharp fluctuations, primarily because of a single product category. Trinidad and Tobago, on the other hand, developed a diverse export mix and experienced a relatively steady rise in export earnings for the entire period examined (Figure 7.9). Unlike in Barbados, where electronics exports were mainly consumer goods, Trinidad and Tobago's electronics growth was in its aircraft and boatbuilding industries. Trinidad's nearly steady growth in chemicals is also the result of a variety of products, and its exports from the manufactures category are iron and steel. Trinidad and Tobago's oil revenues, which were $17 billion during the windfall period 1974-1983, enhanced its ability to invest in infrastructure and attract foreign capital compared with the other Caribbean countries (see Thomas, 1988, p. 279). These oil revenues have been used to promote large-scale joint ventures, involving government-owned operations and foreign investors, in both the chemical and steel industries ("Spoiling the Works," 1984). Its smoother and more steadily upward pattern of industrial exports is in large part attributable to greater industrial investment from domestic public and private sources, such as in its heavy industrial complex at Point Lisas (see Thomas, 1988, pp. 86, 283), which are less prone to withdrawal than investment by MNCs.

Having reviewed the export industrial trends for the Caribbean, we now turn to their possible explanations. Factors operating at the three geographic scales of the world economy, U.S. policy, and Caribbean policy are explored in turn. We will assess the explanatory power of each set of factors through a combination of three criteria: plausibility at the conceptual level, support in the literature, and, most important, ability to account for the temporal and geographic patterns of Caribbean industrial exports (Table 7.1 and Figures 7.1-7.9).

Regimes of Accumulation

To what extent are contemporary Caribbean industrial trends attributable to macroeconomic cycles and, in particular, forces encouraging U.S. firms to relocate plants to accessible sites that offer cheap labor? We address this question by first summarizing the relevant economic theory, then reviewing literature that supports such an interpretation, and finally by considering the theory's fit for trends in industrial exports from the eight Caribbean countries.

A body of evolutionary economic theory describes regular historical cycles of profitability for products and, because products share innovations and production systems, more loosely for whole economic systems. Theorists use several labels for these ideas, including regimes of accumulation, long-wave theory, and profit or product life-cycle theory (Dicken, 1992; Markusen, 1987; Scott & Cooke, 1988; Wilson, 1990). Despite the variety of nomenclature, each theory describes a bell-shaped curve for which the x axis is time and the y is the profit rate. The names and portions of the curve delineated by the theories also differ, but the general idea is that profitability rises quickly and early in the cycle, peaks, falls off, then stagnates, awaiting the emergence of more innovative products and production methods and the beginning of a new profit cycle. The profit cycle produces a geographic pattern of industrial investment with respect to central and peripheral economies. Investment gravitates toward regions of innovation and high profitability, and thereby creates an industrial core. When profits are falling to unacceptable levels, industrial investment disburses toward sites of cheap labor in order to stave off the decline and the competition from more innovative industries. Full economic cycles historically take 40 to 60 years, but recent computer-based innovations seem to have shortened product life cycles considerably (Goss, 1991).

The dominant regime of accumulation in the post-World War II period, Fordism, features long production runs of standardized products for mass consumption and was concentrated at its peak in huge industrial cores such as the North American Manufacturing Belt (Mayer, 1989; Scott, 1988). Measured at the level of the U.S. economy as a whole, the profit rate for Fordism had peaked by 1966 and has declined since (Goss, 1991). Concerning the role of the Caribbean in Fordism, industrial investment is predicted when profits in the core are falling—in other words, after 1966—and should continue until the centripetal forces of innovation and a new regime of accumulation beckon it back to central economies. Disinvestment from peripheral sites such as the Caribbean is then expected in favor of reconsolidation of new industrial regions in the core.

The emerging, post-Fordist regime of accumulation, called flexible or "just-in-time" production, is expected to have dire consequences for peripheral industrial platforms, such as the Caribbean (Dicken, 1992, pp. 117-118; Griffith, 1987, p. 72; Storper, 1991; Wilson, 1990, pp. 140-141). Flexibility derives its profitability from close contact between industries and suppliers, production adjustability, and short production turnaround. These require close coordination of all phases of production. Flexible industrial production is therefore expected to encourage firms headquartered in developed countries to reconsolidate their operations in the most profitable locations in the core. If this change in accumulation regime is real and complete, offshore assembly operations such as those in the Caribbean will no longer be attractive.

Despite abundant literature dichotomizing regimes of accumulation (e.g., Storper, 1991), however, they are not in reality temporally distinct. It is more accurate to understand flexible production as an emerging and competitive production process, rather than as an immediate and wholesale replacement for Fordist firms. The transition is gradual, and there is room in waning Fordism for industrial activity from peripheral sites. Like other observers (e.g., Sadler, 1992), we find it premature to write Fordism's epitaph and declare simultaneously the hegemony of flexible specialization. We remain unconvinced that spatially disbursed Fordist industry is currently being eliminated by more competitive firms in core countries utilizing flexible specialization. We find more accurate the view that the two production systems are not dichotomous and incompatible. They now coexist and are likely to continue to do so well into the future (Mayer, 1989; Preteceille, 1990).

Having reviewed the theory, we can now turn to some illustrations of the range of evidence in favor of such an interpretation of PACE. If the notion of accumulation regimes correctly describes the buildup of Caribbean industry, it should feature the mass production of Fordism more than flexiblism. Reviewing the investment and industrial patterns, Griffith (1987) argues in no uncertain terms that it was the crisis of Fordism that shifted production to the Caribbean. That Caribbean industries are Fordist is also supported by numerous empirical descriptions of Caribbean industrial zones characterized by routinized mass production by low-skilled labor (Deere et al., 1990; Sunshine, 1988). Similarly, the chair of MacGregor Corporation, in the context of discussing the firm's Haitian production facilities, has said that many U.S. firms would fold if they didn't engage in offshore assembly (*Caribbean Report,* 1984, p. 7). In parallel fashion, Hope (1989) observes that firms are driven to the Caribbean by economic forces, and that they need no active policy incentives from Caribbean or other Third World countries to set up operations there. He argues that the decision to move to a country is usually made independent of both host and home country. Policies in the host countries are limited to influencing where within the country the investment will occur and its precise form (Hope, 1989, p. 73).

Goss (1991) makes the strongest case, backed by quantitative evidence, for a regime of accumulation-based interpretation of Caribbean industry. He argues that the U.S. electronics industry moved through a product life cycle in recent decades and responded at the predictable time with industrial investment in the Commonwealth Caribbean, taken as a whole. More specifically, he argues that electronics, like the U.S. economy as a whole, underwent a recession in 1978-1982, which, as the theory predicts, caused a "rationalization and restructuring" of production (p. 39). Electronic investment in the Caribbean should therefore have been higher during the recession than before and after, when the industry was again more profitable and therefore less pressured to seek sites of cheap labor. Goss's statistical results support the theory. They reveal that the annual U.S. imports of electronics products from English-speaking Caribbean countries, taken as a whole, were significantly higher during the recession than in 1960-1977 or 1983-1987. Further support for this interpretation is provided by the fact that U.S. investment across *all* economic sectors in the Caribbean, most notably petroleum, peaked during 1977-1982 and has since declined (Deere et al., 1990, p. 164).

Although Goss's evidence and interpretation provide legitimacy to the theory of regimes of accumulation as an explanatory tool for Caribbean industry, the interpretation is less persuasive across the variety of Caribbean countries examined here, and at the country level, even within the Caribbean Commonwealth. U.S. electronics exports were *also* larger during the 1978-1982 recession (Figure 7.10), and so the core and peripheral regions followed production patterns that parallel, rather than mirror, one another as the theory suggests. In Barbados, the Caribbean's largest per capita participant in electronics, exports in 1983-1986, although they were falling, were still significantly higher than during the 1978-1982 recession (Figure 7.2). Jamaica, the largest English-speaking country, had virtually no electronics exports until 1981, and since then the per capita value has been modest (Figure 7.7). The Dominican Republic had virtually no electronics exports until after 1981 (Figure 7.4). Electronics exports from the other Caribbean countries provide at best modest support for the theory (Figures 7.2-7.10).

Before drawing conclusions, we now consider the explanatory power of regimes of accumulation and kindred theories for the other three industrial export categories, as well as for total industrial exports (Figures 7.1-7.10). If the notion of regimes of accumulation explains Caribbean export industrialization, the recent historical profiles of the countries for each industrial category should have the same peak and trough years. However, as noted for electronics, and as the data indicate for the other industries, countries experienced their peaks and troughs in different years (Table 7.1). Further, if a decline in production from a Caribbean country signifies reinvestment in and reconsolidation of the core, the theory cannot account for a *resurgence* of Caribbean industrial exports after the decline. Some Caribbean countries, the Dominican Republic, for example, do feature peaks and troughs, followed by additional expansion (Figure 7.4). We interpret this as a mismatch in the level of abstraction of the highly generalized theory and the industrial trends of particular countries such as those of the Caribbean (Preteceille, 1990). Theories are of limited use for describing geographic movements of economic sectors, such as electronics or apparel, to each Caribbean country.

From this review of the theoretical and empirically based literature as well as industrial export trends for the Caribbean, we arrive at the following conclusions. Fordist industries in the United States generally underwent a descent of profitability after 1966, and specific industries deviated somewhat from the periodicity of the economy as a whole. Since the late 1960s, a substantial number of U.S. firms coped with decline by shifting labor-intensive assembly operations to Third World sites, including the Caribbean. The theory of regimes of accumulation is most useful in explaining broad economic and regional investment patterns (Figure 7.1), but loses explanatory power at the country and sectoral levels (Figures 7.2-7.9). The geographic outcomes of the evolution of regimes of accumulation are more varied than one might believe from the literature. We now turn to two types of policy-based interpretations of Caribbean industrialization to assess their explanatory contributions.

The Caribbean Basin Initiative

Before examining government initiatives to encourage Caribbean industrial development, we must acknowledge a point, raised by Aglietta (1979), Goss (1991, pp. 12-13), and others, that policy is not independent of the periodicity of regimes of accumulation. It is no coincidence that, beginning in the late 1960s and corresponding with falling profit levels in the U.S. economy, the U.S. government created a series of policies that make the relocation of labor-intensive aspects of production to the Caribbean and other parts of the Third World more profitable for U.S. firms. Having said that, it is also true that policies have some autonomy from economic cycles. For one thing, policies seldom match the needs of firms in a timely way, in part because the ratchetlike buildup of policies does not correspond with economic periodicity. Policies also tend to favor some economic sectors, product components, and countries more than others (Barff & Austen, in press). We examine policy here to gauge its unique contribution to PACE.

CBI is the political acronym given to the Caribbean Basin Economic Recovery Act (CBERA), which became U.S. law in 1983. The stated intent of CBERA was to provide incentives for foreign and domestic investment in nontraditional sectors of Caribbean economies. The U.S. government argued that by boosting these sectors, CBI would help to diversify Caribbean economies and expand their exports (U.S. Department of Commerce, 1991, p. 1). Despite its political prominence, CBERA, as a specific tariff policy, is only one small change in a series of U.S. trade laws extending back to the 1970s and designed to increase the flow of capital and products between the United States and the Caribbean (Deere et al., 1990). No single U.S. policy toward the Caribbean has dramatically altered trade conditions, but together, U.S. policies open U.S. markets to an array of products assembled in the Caribbean. We therefore use the acronym CBI in its broadest sense to refer to all U.S. policies toward the region during the past two decades designed to open new types of trade between the United States and the Caribbean and to increase the region's industrial exports to the United States (Newfarmer, 1986). More specifically, we identify four significant layers of U.S. policy in the last two decades encouraging Caribbean export industrialization: (a) items 807 and 806.30 of the U.S. Tariff Schedules (USTS), (b) the U.S. Generalized System of Preferences (GSP), (c) CBERA, and (d) bilateral textile agreements (807A).

Items 806.30 and 807 became law in 1969 precisely to encourage offshore industrial activity in the areas of metal processing and product assembly, respectively. They allow components to reenter the United States with duty charged only on the value added from processing abroad (Clark, 1989, p. 864; Grunwald & Flamm, 1985, pp. 34-37). Yearly quotas set the limit for each country's participation in these programs. In our Caribbean context, apparel and electronics exports account for almost 90% of exports to the United States under 806.30 and 807 (Clark, 1989, p. 863).

The U.S. GSP, adopted in 1976, grants certain products from developing countries duty-free access to the U.S. market. Originally, GSP was to last for only 10 years, but was extended another 8½ years in 1985. The policy extends its benefits to almost 140 countries having most favored nation (MFN) status, and excludes products, such as steel, glass, textiles, and footwear, that other U.S. laws restrict. To receive the duty exemption, a product must have at least 35% of its value added in the beneficiary country and not exceed $28 million in total value. There are also restrictions on a beneficiary's political behavior. For example, GSP is not available to countries that do not cooperate with U.S. drug-trafficking suppression, or to countries granting trade preferences to other developed countries to the detriment of U.S. trade. When GSP was extended in 1985, a new provision was added that allowed for the exclusion of countries whose products were deemed competitive without aid. Hong Kong, South Korea, Taiwan, and Singapore were thereby declared ineligible for GSP in 1989 (Brown, 1989, p. 347; Kreinin, 1991, pp. 428-429).

The third notable layer of policy, CBERA, provides for the duty-free entry into the United States of a variety of Caribbean products, including industrial goods as diverse as integrated circuits, baseball equipment, and analgesics, while excluding others, such as textiles, apparel, footwear, and watches (Brown, 1989, pp. 339-340; Newfarmer, 1986; Ramsaran, 1989; Schoepfle & Perez-Lopez, 1989). CBERA also provides increased economic assistance for Caribbean governments to use to aid private sector development, and aid to U.S. investors, including financing and support from U.S. government agencies, such as USAID, the Peace Corps, and the U.S. International Trade Commission (Newfarmer, 1986, pp. 67-68; Pentojas Garcia, 1988, pp. 47-48). Like the GSP, CBERA also requires benefiting countries to support U.S. economic and political strategies (Ramsaran, 1989, p. 140; Wilson, 1990, p. 224). Many exports granted tariff exemption by CBERA were already exempt under GSP because the same Caribbean countries have MFN status. Other industrial products, if they have U.S. components, have preference under items 806.3 and 807. Because of these policies that were introduced in the years prior to CBERA, tariff rates on dutiable Caribbean products were already reduced to about 2.5%. However, they continued to be around 25% for textiles, which were also restricted from entering the United States by quotas (Clark, 1989; Deere et al., 1990; Newfarmer, 1986).

In 1986, a provision called 807A or Super 807 was added to the original 807 rule providing for bilateral agreements to increase export quotas for textiles and apparel assembled from components made and cut in the United States (Deere et al., 1990, p. 160; Wilson, 1990, p. 225). By 1989, the Dominican Republic, Haiti, Jamaica, and Trinidad and Tobago signed agreements to participate in 807A (Deere et al., 1990, p. 161; Schoepfle & Perez-Lopez, 1989, p. 157; Wilson, 1990, p. 225).

Our review of U.S. policies promoting PACE leads to two observations. First, CBI not only opens the U.S. market to Caribbean goods, it also, perhaps more important, increases the access of U.S.-based MNCs to the Caribbean's low-wage labor pools. Second, the policy is temporally gradual and quite inclusive of a variety of manufactured goods.

One therefore expects that if U.S. policy has a significant impact on the Caribbean, evidence would take the form of regular, upward linear growth of PACE from the 1970s onward. We now turn to the evidence of the impacts of U.S. policy in the literature, and to the question of whether U.S. policy can account for trends in our data.

Quantitative assessments suggest that CBERA has increased Caribbean industrial exports to the United States, but only by minor amounts. By one estimate, CBERA would increase total exports from the Caribbean Basin to the United States by $61.5 million per annum, with the great majority of the value in manufactured goods (Sawyer & Sprinkle, 1990). However, if the annual monetary impact of CBERA is divided by the Caribbean population, as in Figures 7.1-7.10, CBERA contributes at most only a few additional dollars worth of per capita exports. Similarly, Newfarmer (1986, p. 84) offers predictions of long-term increases in exports owing to CBERA that are under $1 annually per capita for the vast majority of Caribbean Basin countries. There is also evidence of a lack of temporal fit between U.S. policies and PACE. Although the 807 value-added tariff policy extends back to 1969, 807 exports, particularly apparel and textiles, grew faster as a share of total exports than the CBERA contributions from 1985 to 1987 (Deere et al., 1990, p. 167). This suggests that U.S. policy provides a favorable context for industrial development but does not necessarily trigger it in the short term.

Turning to an assessment of whether there is evidence of CBI's impact in Figures 7.2-7.9, we are most struck by how the peaks and troughs of most countries' exports do not follow from increasingly favorable U.S. policy. CBI may provide more profitable opportunities for U.S. industrial investors, but the response has been sporadic with respect to individual Caribbean countries. This mismatch between U.S. policy and industrial exports at the country level suggests that alternative sites for PACE are far more numerous than U.S. firms seeking those sites.

Our assessment is that CBI has only marginal effect on the extent of peripheral investment for industrial export. A greater contributor to Caribbean export growth than U.S. regional policy is U.S. economic expansion in general. By Newfarmer's (1986) calculations, U.S. economic growth or decline has a far greater positive or negative effect, respectively, on Caribbean exports than all specifically export-inducing policies combined. This suggests extreme dependency; Caribbean economies, in classic peripheral form, are appended to that of the United States and vacillate in a more exaggerated way than the ups and downs of the core economy.

Other evidence reveals that the periodicity of U.S. investment in the electronics industry in the Caribbean not only does not correspond with expanded export promotion by the U.S. government, but actually correlates *negatively* with it. In 1984, U.S. semiconductor imports from Barbados soared from just over $20 million in 1982 to $147 million, and represented about half of all of Barbados's exports to the United States. However, electronics were subject to tariff duty throughout this period (Goss, 1991, p. 58). As of March 1985, semiconductor imports from Barbados could enter the United States duty-free, but imports for that year fell to $108 million (Pelzman & Schoepfle, 1988, p. 763). By 1987, such imports were below their 1982 level (Figure 7.2; Goss, 1991, p. 58). In this example, CBI is ineffective at increasing investment in and exports from Barbados.

A possible explanation for the remarkable decline of Barbados's electronics industry during CBI is that the U.S. policy is anachronistic when compared with regimes of accumulation. In other words, by the late 1980s U.S. electronics industries no longer were attracted to cheap labor, but instead were reconsolidating investments in the United States (Goss, 1991; Griffith, 1987, p. 72). If this is true, it offers a more significant challenge to the unique contribution of tariff policies than was implied when we argued that they are not independent of regimes of accumulation: Not only can policy be viewed as a stopgap response to economic crisis; in some contexts, the effects of policy are actually irrelevant to industrial trends.

Another interpretation of the decline in Barbados's electronics and other industrial exports since 1983 (Figure 7.2) is that state-indexed wages in Barbados have risen to be uncompetitive with other Caribbean countries that were more aggressively seeking foreign investment and devaluing their currencies in the 1980s. If true, this too would suggest that CBI cannot take credit for growth in PACE at the country level. Wage policies are discussed below, as they are a component of local industrial policy activism.

Export Processing Zones

If it can be argued that U.S. policy favoring Caribbean industrial exports is a response to economic demands for access to sites with low-cost labor (Griffith, 1987), then the policies of Caribbean countries are even more closely tied to economic stimulus. Caribbean industrial policy is oriented heavily toward the development of EPZs and their benefits for foreign investors. In this section we address the question of the extent to which local policy activism matters in the explanation of the international geography of industrial production. In other words, to what extent have Caribbean countries fashioned their own industrial development policies that shape the quantity and quality of local investment and production?

As discussed above, Caribbean countries have been motivated by external factors to pursue development strategies based on exports. A key to this strategy has been to lure MNCs to Caribbean sites to conduct assembly operations. The major vehicles for this are EPZs, which are usually located near ports or international airports and are accessible to ample labor forces willing to work for near subsistence wages. Most PACE activity takes place in EPZs, which can be found in virtually all Caribbean countries. Host governments consider EPZs to be outside of their national customs territories (Schoepfle & Perez-Lopez, 1989, pp. 131-133). Firms are free to conduct their business however they like within the zones: They can import equipment and raw materials without paying duty as long as goods leave without entering the host's official customs territory, although some countries also allow producers to sell some of their finished products in local markets. Caribbean governments give investors in the zones an exemption, lasting from 10 to 15 years, from income taxes, as well as subsidized utilities and factory shell rentals (Schoepfle & Perez-Lopez, 1989, p. 134).

The earliest Caribbean EPZs appeared in the 1960s. Haiti established the first Caribbean zone in 1960, and after that the number of firms operating in Haitian EPZs expanded, from 13 in 1966 to 154 in 1981. Assembly operations in Haiti produce apparel, sporting goods (mostly baseballs), and toys. The Dominican Republic established its first EPZ in 1969, and by 1986 it had 5 EPZs, with another 6 being built and 12 under development. Jamaica had 3 EPZs by 1986, with most space devoted to apparel assembly. Barbados has targeted electronics, medical supplies, high-quality apparel, and data processing for its EPZ development; in 1985, there were 23 firms in Barbados producing these goods (Deere et al., 1990, p. 145; Schoepfle & Perez-Lopez, 1989, pp. 137-139). Trinidad and Tobago has only recently developed EPZs per se, but has lured foreign industries with benefits similar to those offered by its Caribbean neighbors.

The fact that Caribbean EPZs and the multiple layers of CBI emerged and expanded their benefits to foreign investors in PACE since the late

1960s is revealing. These policies correspond with falling profits in the U.S. economy and the search by industrial firms for lower labor costs. Both U.S. and Caribbean policy appear to respond to demands in industry, rather than to represent an autonomous initiative on the part of policy makers to construct new economic relations between the regions.

Nonetheless, some have argued that there are important contributions from, and variations in, local policy that are critical to industrial expansion in the Caribbean. Deere and her collaborators (1990), for example, having considered the very low correlation between growth in Caribbean investment and export in the 1980s and U.S. policy designed to stimulate it after 1983, conclude that local government policy explains industrial patterns: "A greater stimulus to investment and employment generation are the conditions offered by the free-trade zones themselves: cheap and docile female labor and attractive government policies" (p. 172). Quite arguably, U.S. firms would invest in other regions if Caribbean governments did not make their PACE sites increasingly attractive. And, increasingly it seems, in the context of greater ease of movement for capital and products that tariff reductions and EPZs provide, investors discriminate among countries in the Caribbean Basin regarding the relative cheapness, docility, and stability of their labor pools. We next review some examples of this trend, in which the critical local policy encouraging PACE reduces to the suppression of labor demands.

Until the mid-1980s, the Dominican Republic's industrial export output had not been impressive when gauged in per capita terms, nor had it been upwardly linear over time (Figure 7.4). However, since then it has greatly expanded its EPZ sites, has increased EPZ employment to 100,000, and has been held out as an example of how local policy can effectively distinguish a country from its competitors (Hillcoat & Quenan, 1991). Part of the explanation of the growth of PACE in the Dominican Republic comes from the country's industrialization policies of the past two decades. These include its relatively extensive ISI program based on foreign investment between 1966 and 1976, and its state development program, called CEPODEX, specifically designed to develop nontraditional exports since the late 1970s. Under these policies, the Dominican Republic developed the infrastructure and ties to foreign capital necessary for export industrialization (Hillcoat & Quenan, 1991, pp. 213-215).

On the other hand, two labor-related factors are more useful for explaining the *timing* of expansion of export assembly operations in the

Dominican Republic since the mid-1980s. One is Haiti's recent anti-oligarchic political turbulence, leading to the fall of the Duvalier dictatorship in 1986 and the election of progressive, and since deposed, President Jean Bertraud Aristide in 1990. Until 1985, Haiti had the region's lowest wages and the most firms operating in EPZs, but Haiti's popular insurrection has made it much less attractive for foreign investors (Hillcoat & Quenan, 1991, p. 215). Still, of all other Caribbean investment platforms besides Haiti, why is the Dominican Republic so attractive? Labor costs are key. Currency devaluation has made the Dominican Republic's minimum hourly wage (U.S. $.55 in 1988) the lowest labor cost in the Caribbean (Deere et al., 1990, p. 149).

Barbados's trajectory for wages directly contrasts with that of the Dominican Republic, and its experience with PACE is also in the opposite direction. From the late 1960s until 1980, Barbados's industrial exports grew by 28% per year. By the early 1980s, Barbados was by far the largest per capita industrial exporter among Caribbean countries, with electronics and garments contributing the most (Figure 7.2). During this period of industrial export expansion, and unlike all other Caribbean countries except Trinidad and Tobago, the Barbadian state regularly increased the wage index so that it nearly kept up with consumer price inflation. At the same time, the Barbadian state did not severely devalue the currency against the U.S. dollar (Thomas, 1988, pp. 272-275). Since 1983, however, exports have plummeted. Deere et al. (1990, p. 179) argue that mounting tariff advantages for U.S. industries to produce in the Caribbean have removed Barbados's earlier, more market-based, attractiveness to U.S. firms. The $2.16 per hour paid to Barbadian semiskilled workers in EPZs by 1988 was uncompetitive with the many other Caribbean EPZs, where workers receive less than $1.00 per hour (Deere et al., 1990, p. 149). In the context of expanding opportunities for PACE across the region, local policies that allow for rising working-class wages, as in Barbados, make the country unattractive to foreign industrialists.

Another factor explaining the irregular and scattered pattern of PACE across Caribbean countries (Figures 7.2-7.9) is that, despite pro-investment policies established by the governments of the United States and the islands, U.S. investors perceive most Caribbean countries to be less than secure and stable investment environments. Even for the Caribbean countries with high investment security ratings in the early 1980s, Dominica and St. Lucia (Newfarmer, 1986, p. 70), exports vacillate. Further, the limited-term tax holidays that each Caribbean government offers MNCs to locate assembly plants on its island inadvertently decrease the firms' attachment to the location and its local dependence and thus encourage hypermobility. Caribbean policies increasingly have allowed unrestricted repatriation of profits, and this encourages foreign firms to reinvest where profit making is highest. Together with low capital and job training investments, Caribbean policies make foreign industries "particularly footloose" (Deere et al., 1990, p. 181), contrary to the islands' own long-term interests. Thomas (1988), for example, observes that firms operating in Caribbean EPZs "are usually able to close down and move elsewhere in a matter of days" (p. 94). In this case, "elsewhere" is often another Caribbean country's EPZ, where labor is less organized or cheaper, or where policies offer the firm greater infrastructural or tax benefits. Thus the flexibility in Caribbean export industries is enhanced capital mobility, to the distinct disadvantage of each island economy. In other cases, relocation involves abandoning the Caribbean completely to relocate in the United States and to participate in flexible production networks (Storper, 1991).

Local industrial policy matters, although not necessarily in a way that signals a new and more autonomous localism in the Caribbean. Understanding why more fully requires that we appreciate the extent to which local policy is actually shaped by core interests. It is no mere coincidence that EPZs and kindred policies promoting PACE emerged in all Caribbean countries under investigation here during the past two decades. True, EPZs are created by local policy, but to a significant extent they are local reactions to pressures from global political and economic actors. In the economic sphere, EPZs can be viewed as smaller-scale and concentrated versions of regional trading blocs. Multinational firms have pressured governments to join regional trading blocs so as to remove nationalistic barriers and restrictions on foreign investor behavior (North American Congress on Latin America, 1991). EPZs offer foreign investors much more flexibility and subsidy than found in any free trade agreement, but the benefits apply only to industrially related activities within the zones, rather than countrywide. In the international political arena, EPZs are responses to pressure from the IMF and other core state interests on heavily indebted Third World governments to open opportunities for foreign investment and exports. Either way, the new localism of industrial export policies in the Caribbean puts local governments into hypercompetition with one another, with lower wages, larger subsidies, and greater freedom in the movement of capital offered to attract firms. These policies inhibit the emergence of a more independent and creative new localism.

Caribbean Industrial Exports: New Localism or Old Dependency?

As a region, the Caribbean has expanded and diversified its industrial exports impressively since the 1970s. Several factors have converged to fuel the industrial surge. U.S. manufacturers have sought low-wage sites for product assembly. The U.S. government has created several layers of tariff policy encouraging industrial investment in and export from the Caribbean. And virtually every Caribbean country now has an active policy to attract foreign manufacturing investment and offers one or more free trade zones with special legal status for encouraging foreign investment and export production. Despite the convergence of factors encouraging growth, policies and market uncertainties offer few incentives to foreign firms to establish themselves for the long term in specific Caribbean EPZs. To the contrary, foreign firms have demonstrated in recent years how they can invest, move profits, and withdraw operations from particular sites with remarkable ease. Thus foreign investment for industrial exports has taken most Caribbean countries for a very uneven ride.

In this chapter we have investigated the extent to which recent industrial developments in the Caribbean are attributable to three scale-dependent processes: (a) macroeconomic trends explained by the concepts of long waves, regimes of accumulation, and profit cycles; (b) U.S. policy toward industrializing, economically diversifying, and integrating with the region; and (c) the export industry-promoting policies of individual Caribbean countries. Observing trends in industrial exports for the Caribbean as a whole, we find that each process contributes to the explanation distinctively from its own geographic niche in the international political economy, and that the explanatory power decreases markedly from the first to the second to the third process. In other words, economic processes dictate and political policies respond to them; the policies of core countries are more influential than those of peripheral countries. Local policy in the Caribbean is essentially reactive rather than proactive. Local policy responds to pressures from global economic restructuring and from larger and more powerful interests, and is redundant and hypercompetitive among the islands.

This movement across the islands to create opportunities for profitable investment is an extreme and rather pitiful version of the trend in advanced capitalist countries for local government to be more "entrepreneurial" in its pursuit of economic development (Harvey, 1989; Mayer, 1989; Preteceille, 1990).

When we move to the more specific level of explaining industrial export patterns for individual Caribbean countries, the same pattern of relative explanatory power applies, although the explanatory power of each theoretical perspective at the country level is substantially less. We observe only a weak correlation between the propulsion of labor-intensive industry toward accessible Third World sites such as the Caribbean and actual investment and export production on specific Caribbean islands. In fact, neither of the types of explanation can account for most of the empirical trends for individual countries. We must therefore consider what produces the *lack* of fit among economic forces, firm behavior, and industrial outcome per location. We understand this to mean that although macroeconomic trends put pressures on firms, they do not dictate any specific response. Among their many options for contending with a falling rate of profit, firms can attempt to extract wage concessions from their current workforces in the core, introduce new competitive technology, close down and move the capital to another sector, or relocate production to a lower-wage site. Potential lower-wage sites are many more than potential investors, leaving the vast majority of low-wage sites ripe for investment but without it (Storper, 1991, chap. 8). In sum, falling profits among core industrial firms and their effort to reduce costs by no means produce a singular and specific geographic outcome.

The rise and fall of industrial exports from most Caribbean countries are related to a broader characteristic of the world economy. The supply of countries courting investment and offering profitable production sites far exceeds the core firms' demand for them. At the same time, however, Caribbean—and, for that matter, Third World—export industrialization is not simply randomly scattered across territories, but instead is geographically clustered in a few locations. The clustering can be explained largely by policy initiatives of core and, secondarily, peripheral countries. Development policies of core countries such as the United States that favor certain regions or countries, as has been the case with the East Asian NICs, Mexico, and Puerto Rico, channel investment and industrial expansion to these countries. No matter how creative Jamaica, for example, is with its own industrialization policy, it is unlikely to match nations targeted with particularly favorable policies from the core. Among Caribbean countries operating under similar rules from the core, local policy makes some difference. Countries with the earliest and most sustained policies to lure foreign industrial investment, such as the Dominican Republic, have built reputations, developed experienced workforces, and gathered enough industrial activities to generate a local context attractive to industry over the longer term. Trinidad and Tobago's industrial development experience is similar to that of the Dominican Republic, but with the added advantage of massive oil profits to pay for industrial infrastructure, which no other country in the region can match. Less well endowed and more recent joiners in the pursuit of foreign industrial investment, such as St. Lucia and Dominica, have been given an especially temporally uneven introduction by footloose firms.

Regarding the component of the new localism that involves community groups in the creation of policy, we find remarkably little evidence of broad interest representation within Caribbean countries in their pursuit of foreign investment and industrial exports. The impact of export industrialization is even less positive for the local working class than for Caribbean countries as a whole. This is because industry largely employs young females, who acquire few skills during their temporary employment. Decisions made by political elites are availed to them by forces outside their control and by interests more powerful than they are.

We conclude by asking, Who controls and therefore benefits from Caribbean industrialization? The Caribbean region's greater incorporation into an international division of labor does not decrease its dependence on the United States. Although it is possible that economic diversification and expanded industrial bases may eventually offer CBI countries a few more development options, any industrial policy so dependent on the policies and decisions of external powers, like the Caribbean's EPZs, is bound to be fragile. Poor people in the Caribbean are most likely to benefit from economic policy that is genuinely local in control. Caribbean government regimes must answer to some extent to the local working class, and local populations have ousted the most undemocratic ones. On the other hand, the investors and politicians from outside the region that play so large a role in Caribbean export industrialization are in no way accountable to Caribbean populations. As we have argued in this chapter, Caribbean countries do not control the industrial export development policies that their leaders enunciate. Caribbean governments have engaged in *local political activism* to lure foreign industries, but this activism does not signify more local democracy and autonomy. To the contrary, the activism responds to outside dictates and further deepens the islands' position on the fringe of the global economy. And, given the continuing ability of the IMF and other interests from the core to shape island policies, it seems unlikely that Caribbean governments will have sufficient autonomy to construct markedly different development policies in the near future.

Notes

1. In the summer of 1992, the United States, Canada, and Mexico signed the North American Free Trade Association treaty aimed at eliminating their investment and trade barriers. U.S. policy makers have proposed expanding the free trade zone to encompass all of Latin America and the Caribbean.

2. We exclude Puerto Rico, an autonomous commonwealth of the United States, which receives huge annual subsidies from the United States, and from which it has developed much greater industrial capacity and diversity (Deere et al., 1990, p. 37).

3. With CBI, the Reagan administration sought to award compliant countries and to punish mavericks. Cuba, Nicaragua, Guyana, Suriname, the Cayman Islands, the Turks and Caicos Islands, and Anguilla were excluded from the U.S. policy. Nicaragua and Guyana were allowed in later, when right-wing governments came to power (Deere et al., 1990, p. 156).

4. Some countries did not report industrial exports for certain years, and 1988 is the most recent year for which comparable statistics are currently available. For years in which no cumulative data were given for a particular broad grouping, totals were estimated by adding up all listed subcategories. An emergent industry in the Caribbean is information processing, which takes forms such as computer data entry for U.S. firms. Traditional industrial data sources such as the *United Nations International Trade Statistics Yearbook* report only tangible products, thus new data sources are needed to capture information-based activity.

5. The growth is in part from investments by the MacGregor Corporation, which started manufacturing baseballs and softballs in Haiti in 1978. MacGregor had hired 1,000 workers by 1983 and had plans to double that number in 1984. In the 1980s, Haiti's attractions included the lowest Caribbean wage rate, a productive workforce, and a group of nationals who were Canadian- or U.S.-trained businessmen. By 1984, the U.S. ambassador to Haiti said that it had achieved peace and stability, and he praised Haiti's compliance with the IMF austerity program, a bilateral investment treaty with the United States, and its judicial and labor reforms of the past 10 years (*Caribbean Report,* 1984).

References

Aglietta, M. (1979). *A theory of capitalist regulation.* London: New Left.

Barff, R., & Austen, J. (in press). "Its gotta be da shoes": Domestic manufacturing, international subcontracting and the production of athletic footwear. *Environment and Planning A.*

Brown, D. K. (1989). Trade preferences for developing countries: A survey of results. *Journal of Development Studies, 24,* 335-363.

Caribbean Report (Latin American Regional Reports). (1984, June 15). Haiti. RC-84-05, p. 2.

Clark, D. P. (1989). Measurement of trade concentration under the United States' Caribbean Basin Initiative. *World Development, 17,* 861-865.

Codrington, H., & Worrell, D. (1989). Trade and economic growth in small developing economies: Research on the Caribbean. In D. Worrell & C. Bourne (Eds.), *Economic adjustment policies for small nations: Theory and experience in the English-speaking Caribbean* (pp. 28-47). New York: Praeger.

Deere, C. D., Antrobus, P., Bolles, L., Melendez, E., Phillips, P., Rivera, M., & Safa, H. (1990). *In the shadows of the sun: Caribbean development alternatives and U.S. policy, policy alternatives for the Caribbean and Central America.* Boulder, CO: Westview.

Development: A guide to the ruins [Special issue devoted to the writings of Wolfgang Sachs]. (1992, June). *New Internationalist, 232.*

Dicken, P. (1992). *Global shift: The internationalization of economic activity.* New York: Guilford.

Golub, S. (1991). The political economy of the Latin American debt crisis. *Latin American Research Review, 26,* 175-215.

Gordon, D. (1988). The global economy: New edifice or crumbling foundations? *New Left Review, 172,* 24-64.

Goss, B. (1991). *Offshore industrial location: United States electronics manufacturing in the Caribbean.* Unpublished master's thesis, Indiana University, Department of Geography.

Griffith, W. H. (1987). Can CARICOM countries replicate the Singapore experience? *Journal of Development Studies, 24,* 60-82.

Griffith, W. H. (1990). CARICOM countries and appropriate technology. *World Development, 18,* 845-858.

Grunwald, J., & Flamm, K. (1985). *The global factory: Foreign assembly in international trade.* Washington, DC: Brookings Institution.

Harvey, D. (1989). From managerialism to entrepreneurialism: The transformation in urban governance in late capitalism. *Geografiska Annaler, 71B,* 3-17.

Hey, J. (in press). Foreign policy options under dependence: A theoretical evaluation with evidence from Ecuador. *Journal of Latin American Studies.*

Hillcoat, G., & Quenan, C. (1991). International restructuring and re-specialization in the Caribbean. *Caribbean Studies, 24,* 191-222.

Hope, K. R. (1989). Private direct investment and development policy in the Caribbean: Nationalism and nationalization scared away foreign investors but Reagan's initiative's luring them. *American Journal of Economics and Sociology, 58,* 69-78.

Kay, C. (1989). *Latin American theories of development and underdevelopment.* New York: Routledge.

Kreinin, M. E. (1991). *International economics: A policy approach.* New York: Harcourt Brace Jovanovich.

Lawson, V. (1992). Industrial subcontracting and employment forms in Latin America: A framework for contextual analysis. *Progress in Human Geography, 16,* 1-23.

Logan, J. R., & Swanstrom, T. (Eds.). (1990). *Beyond the city limits: Urban policy and economic restructuring in comparative perspective.* Philadelphia: Temple University Press.

Markusen, A. (1987). Industrial restructuring and regional politics. In R. Beauregard (Ed.), *Economic restructuring and political responses* (pp. 115-143). Newbury Park, CA: Sage.

Mayer, M. (1989, September). *Local politics: From administration to management.* Paper prepared for the Cardiff Symposium on Regulation, Innovation and Spatial Development, University of Wales.

McKee, D. L., & Tisdell, C. (1990). *Developmental issues in small island economies.* New York: Praeger.

North American Congress on Latin America. (1991). The new gospel: North American free trade [Special issue]. *NACLA Report on the Americas, 24*(6).

Newfarmer, R. (1986). Economic policy toward the Caribbean Basin: The balance sheet. *Journal of Interamerican Studies and World Affairs, 27,* 63-90.

Offe, C. (1975). The theory of the capitalist state and the problem of policy formation. In L. Lindberg et al. (Eds.), *Stress and contradiction in modern capitalism* (pp. 124-143). Lexington, MA: D. C. Heath.

Pelzman, J., & Schoepfle, G. (1988). The impact of the Caribbean Basin Economic Recovery Act on Caribbean nations' exports and development. *Economic Development and Cultural Change, 36,* 753-796.

Pentojas Garcia, E. (1988). The CBI and economic restructuring in the Caribbean and Central America. *Caribbean Studies, 20,* 46-58.

Population Reference Bureau. (1992). *World population data sheet.* Washington, DC: Author.

Preteceille, E. (1990). Political paradoxes of urban restructuring: Globalization of the economy and localization of politics? In J. R. Logan & T. Swanstrom (Eds.), *Beyond the city limits: Urban policy and economic restructuring in comparative perspective* (pp. 27-59). Philadelphia: Temple University Press.

Ramsaran, R. (1989). *The commonwealth Caribbean in the world economy.* London: Macmillan.

Sadler, D. (1992). *The global region: Production, state policies and uneven development.* New York: Pergamon.

Safa, H. I. (1986). Urbanization, the informal economy and state policy in Latin America. *Urban Anthropology, 15,* 135-163.

Sawyer, W. C., & Sprinkle, R. (1990). The impact of the Caribbean Basin Economic Recovery Act on Caribbean nations' exports and development: A comment on Pelzman and Schoepfle's estimates. *Economic Development and Cultural Change, 38,* 845-854.

Schoepfle, G. K., & Perez-Lopez, J. F. (1989). Export assembly operations in Mexico and the Caribbean. *Journal of Interamerican Studies and World Affairs, 31,* 131-161.

Scott, A. (1988). *Metropolis: From division of labor to urban form.* Berkeley: University of California Press.

Scott, A., & Cooke, P. (Eds.). (1988). The new geography and sociology of production [Special issue]. *Environment and Planning D: Society and Space, 6,* 241-370.

Spoiling the works. (1984, November 3). *Economist, 293,* 72.

Storper, M. (1991). *Industrialization, economic development and the regional question in the Third World: From import substitution to flexible production.* London: Pion.

Storper, M., & Walker, R. (1989). *The capitalist imperative: Territory, technology, and industrial growth.* Oxford: Basil Blackwell.

Sunshine, C. (1988). *The Caribbean: Survival, struggle, and sovereignty* (2nd ed.). Boston: South End.

Thomas, C. (1988). *The poor and powerless: Economic policy and change in the Caribbean.* New York: Monthly Review Press.

United Nations. (1980). *International trade statistics yearbook.* New York: Author.

United Nations. (1984). *International trade statistics yearbook.* New York: Author.

United Nations. (1987). *Statistical yearbook 1987.* New York: Author.

United Nations. (1989). *International trade statistics yearbook.* New York: Author.

United Nations. (1990). *International trade statistics yearbook.* New York: Author.

U.S. Department of Commerce. (1991). *Guidebook: Caribbean Basin Initiative.* Washington, DC: Government Printing Office.

West, R. C., & Augelli, J. P. (1989). *Middle America: Its lands and peoples* (3rd ed.). Englewood Cliffs, NJ: Prentice Hall.

Wilson, P. A. (1990). The new Maquiladoras: Flexible production in low wage regions. In K. Fatemi (Ed.), *The Maquiladora industry: Solution or problem?* (pp. 1137-1151). New York: Praeger.

8

Local Institutions and Development

The Nigerian Experience

DELE OLOWU

The past decade has witnessed a resurgence of interest in the possible contributions that ordinary people and their locally based institutions can make to national economic and social development. Before this time, the debate seemed to have been settled in favor of the central state. For new nations that had just emerged from colonial rule, the case for centralization was strong. First, the powers of the central state and its key organs, the bureaucracy and the single party, were regarded as important for helping to integrate extant ethnic groups who had been brought together by the accident of colonial conquest. Decentralization, it was argued, can make sense only where there have been strong currents toward centralization in some historical past. Second, because economic development was defined simplistically as the transference of Western capital and technology to preindustrial societies, the only reliable institution to facilitate economic development (as defined above) was the central government and its key organs (for an overview of the literature, see Hyden, 1986; Wunsch & Olowu, 1990).

However, the very disappointing performance of the African centralized states in the first three decades of independence has led to a serious rethinking of the centralized strategy. Today, African national governments as well as donor agencies have indicated some interest in exploring the potential held by the various local institutions (including formal structures such as local governments and informal ones such as community development associations, town unions, and other voluntary associations) for Africa's development. One reason for this interest is the expectation that these institutions can help to mobilize much-needed fiscal (and at times human) resources that could be used for the initiation and maintenance of Africa's declining rural infrastructures, either by themselves or by complementing central government programs. A second reason relates to the lessons that the success of some of these institutions hold for Africa's future in constructing indigenous and workable institutions, especially given the preponderant failure of formally organized structures (Ake, 1987; Esman & Uphoff, 1984; Olowu & Smoke, 1992).

This chapter discusses the Nigerian experience with revitalizing its official grassroots organizations and underscores the significant contributions of nonofficial/indigenous local organizations to the country's economic growth and development. In 1976, a major reform of the system of local government was undertaken in Nigeria, with the primary objective of promoting popular participation, the effective delivery of community services, and the mobilization of human and material resources for community development. A series of postreform efforts have been made to institutionalize these local governments created "from above." Although the 1976 initiative and subsequent incremental reforms have had the effect of strengthening these newly created local institutions, the regularity of change has also served to make them unstable and therefore not as useful for community development purposes.

There also exists another distinct unit of local government in each community, village, or town in Nigeria: the community development agency (CDA). CDAs are nonofficial, generally small in size, and relatively stable in terms of their constitution and level of institutionalization (Awa, 1988; Guyer, 1991; Smock, 1971). It is only recently that national policy makers have begun to pay serious attention to this type of local institution. Based on recent research findings, I argue that the reform of official local governments represents efforts to rationalize central/local political relationships, whereas the nongovernment organizations, being more versatile and sensitive to market pressures, represent attempts by local people to respond collectively to market forces.

The Context: Polity and the Economy at the National and Local Levels

Nigeria is Africa's most populous and only federal state. Even though no precise population figures on Nigeria are available, most estimates hover around 110-120 million. However, a recent but controversial census count put the population of Nigeria at 88.5 million.[1] Whichever of these figures one regards as accurate, Nigeria has a population that approximates roughly a quarter of the black African population. Most Nigerians are to be found in the rural areas. Some 70% live in the rural countryside, and about the same proportion of the labor force is engaged in agriculture and agriculture-related activities. There are, however, a number of very large cities, such as Lagos (estimated at 8 million in 1990), Ibadan (4.7 million), and Kano (2.6 million), besides a number of other secondary cities that now serve as state capitals in most of the 30 states. The national capital was recently moved from Lagos, on the coast, to Abuja, a more central, hinterland location, owing to considerations of space, centrality, and strategy, among others.

Nigeria has a three-tiered government structure consisting of federal, state, and local levels. There are 30 states and 589 local governments, excluding the mayoralty of Abuja, which has four mayoral districts. The federal system has been restructured periodically since 1966, when the military took power from civilian politicians. The uneven structure of four regions was reorganized first into 12 states (1967), 19 states (1976), 21 states (1989), and then into the current 30 states in 1991. States, according to the provisional census figures of 1991, have populations ranging from 5.7 million (Lagos) to 1.4 million (Yobe). As noted above, local governments were reformed nationally in 1976 as part of the efforts of past military governments to correct government structure, promote popular participation, and ensure that Nigeria's windfalls from oil (especially after the Arab oil embargo in 1973) were utilized to provide basic infrastructure throughout the country. A total of 301 officially recognized local governments were created in 1976 out of the diverse regionally based local government structures. An attempt was made by the civilian administration (1979-1983) to increase the number of formal local governments by doubling their numbers or even tripling them in some states, but these were reversed in 1984. In May and August

1991, new local governments were carved out of the original 301 to bring the current total to 589.[2]

The whole of the government system rests on oil, which provides more than 95% of public sector resources and about the same proportion of foreign exchange earnings. Oil has become the lifeline of all Nigerian governments, and in the boom years it resulted in a relentless expansion of all levels of government in terms of scope, agency numbers, and personnel growth. A number of social (health and education at all levels), physical (water, roads, and so on), and economic infrastructures were created as a result, but there were severe costs as well. Nigeria suffered from the "Dutch disease," the near paralysis of all economic sectors except the oil sector. In particular, agriculture, which is the major employer of labor, declined as cheap food imports became readily available because of currency overvaluation and massive private sector borrowing to shore up the predictable oscillations of the oil market. When the oil market collapsed in 1986 and the price of oil fell from its high levels of $43 per barrel to only $6, the country had few options other than to embark on a World Bank-sponsored structural adjustment program (SAP) with the key objectives of stabilization, reduction of deficit subsidies, liberalization of foreign exchange, imports, and government rationalization. This program, initially regarded as short-term, to last only two years (1986-1988), is now in its seventh year. It is currently perceived as a complement to a five-year program of political transition from military to civil rule, which was scheduled to terminate with the inauguration of a civilian president in January 1993. These twin programs of economic liberalization and political transition to civil government have led to another round of institutional reforms, including the democratization and decentralization of local governments.

The SAP has had both positive and negative impacts on Nigeria's rural economy. It has led to momentous increases in agricultural prices and productivity. The rapid increase in the prices of both cash and staple crops led to rapid transformation of the rural countryside; the government consolidated these gains by instituting programs of cheap credit (with the creation of low-income people's banks) and a highly focused program of rural infrastructure construction and rehabilitation through an agency known as the Directorate of Food, Roads and Rural Infrastructure (DFRRI).

There have been pains as well, of course. Because most economic activities in the agricultural and industrial sectors are so heavily dependent on imports, production costs have risen, and this has led to cost-push inflation, low profits, and low capacity utilization. Unemployment has also been on the increase, and the impact of the SAP on the poor, especially those in the urban areas, has been severe. Most important, the country's external indebtedness has increased, rising from $2.3 billion in 1988 to $33.4 billion in 1991.

Local Governments: Political Adjustment of Central-Local Relations

Nigerian local governments owe much to their colonial legacy. Colonial rule was essentially rule by administrators. Because of the small number of British administrators and the large size of the country, a system known as "indirect rule," which incorporates preexisting local chieftains into the modern colonial administrative machinery, was adopted. Even when a federal system of government was adopted in 1948, local governments (referred to as *native authorities*) were responsible for law and order and undertook some development functions (schools, dispensaries, markets, and local feeder roads). In addition, they had local police forces, prisons, and treasuries, together with rudimentary local taxation systems based on head tax, or cattle heads. These native authorities provided most of the domestic services, especially in the northern parts of the country. Their performance led a British student of Nigerian politics to note in the late 1950s that local governments were the "real" governments in the northern region, whereas the regional (state) governments were becoming more important in the southern parts (Hicks, 1961). Even then, local government expenditure as a proportion of national public expenditure hovered between 11% and 15% from 1955 to 1968.

As regional governments began to assert themselves, they reformed local governments in two directions. First, they democratized them by making them elective, but, second, they weakened them further by taking over their responsibilities, revenue sources assigned to them, and even their personnel. Civilian politicians in the regions (states) particularly disliked democratically elected local governments because pockets of opposition emerged in the larger cities (such as Ibadan, Lagos, Kano) to the de facto single-party regimes they had imposed in each of the three regions. Unfortunately, even when the military intervened, the emasculation of democratically elected, semiautonomous local government persisted. In fact, by the middle of the 1970s, 4 of the country's

12 state governments had gone ahead and abolished their local governments altogether and substituted "local administrations" that were run as extensions of the state government bureaucracy. Even where such extreme measures had not been taken, the results were the same: Local governments were devoid of any major responsibilities, tax powers, or other resources, and had only poor quality staff except for the few who were seconded from the state government bureaucracy. Where local governments were allowed to exist as semiautonomous institutions, they had lost any pretense of autonomy. Council members were usually selected by the state governor rather than elected; their personnel, even though separate from the state civil service, were run by agencies set up and managed by the state. Local governments had no revenue sources of their own; rather, they were "assigned" a portion of the state/regional government revenue sources. What was assigned to local governments was subject to continuous and highly irregular reviews carried out unilaterally by the state/regional governments. Local government responsibilities were gradually taken over by the state governments, either directly or by setting up state-controlled parastatals to deliver services they regarded as beyond the ken of local authorities, such as police, water supply, secondary health care, schools, and town planning. Most important, local government systems were in a constant state of flux in terms of structures, services, responsibilities, funding, and resource arrangements.

In 1976, as part of the first transition program to civil rule by the military leadership, a nationwide reform of the local government system was embarked upon. The authors of the 1976 local government reform seemed to have captured fully the pre-1976 situation when they asserted: "Local Governments have over the years suffered from the continuous whittling down of their powers. The State Governments have continued to encroach upon what would normally have been the exclusive preserves of Local Government" (Nigeria, 1976, p. i). Couched in the form of policy proposals, each state governor (also a member of the military) was required to pass identical legislation to enforce the national government's major provisions, including the definition of local governments as democratic units, an articulation of their functions, a financing plan that included the transfer of revenues from the federal and state governments, their organizational structure, and so on.

Implications of the Reform of 1976 for Central-Local Relations

Although it is possible to identify differences among different military administrations, the military has tended generally to be strident in its support for local government revitalization, compared with its civilian counterparts. This is best explained in terms of the workings of the Nigerian federal system (Olowu, 1988, 1990b). Under civilian administrations, state governments are more assertive in the intergovernmental relations arena, whereas under military rule, they operate as extensions of the federal government. Local governments then become the only civilian-based institutions and, in the eyes of the military political leadership, represent the real contact with a restive populace and, most important, with traditional chieftains. Progressively, the military (which has ruled the country longer than the civilians) has proceeded to redefine central-local relations by weakening the state governments (by depriving them of constitutional responsibilities and independent taxes and other revenue sources, such as the units commodity boards, and generally reducing their capabilities through the creation of more states) while increasing the potential strength and vitality of local governments by increasing their responsibilities in proportion to centrally collected resources allocated to them and protecting them from encroachments by the state through constitutional enactments and executive actions. It is not accidental that demands for the abolition of state governments altogether have increased under the present transition program in preference for a unitary state comprising a strong center and local governments (Nigeria, 1986).

Under military rulers, state governments lose their capacity for independent action. Their chief executives (governors) are appointed by the military leadership and are changed at will. The constitution has been amended to ensure that state governments do not take action, even on residual matters, without referring them to the federal government. Even their annual budgets (as well as their five-year development plans) must be approved by the national government. Indeed, in constituting their cabinets, state governors must clear the names of people they intend to appoint with the national government. To all intents and purposes, state governments are run as prefectures.

Furthermore, the responsibilities and revenue sources of state governments have been progressively whittled down under military rule.

First, state units have been increased progressively from 4 in 1966 to 30 at this time—a fact that reduces the resources available; this instability detracts from their stature as institutions capable of autonomous action in the eyes of the public. Second, a number of responsibilities earlier performed by state governments (regulation of commodity boards, police, prisons, local governments, housing, higher education, and the like) have either been completely taken over by the federal government or put on the concurrent legislative list. Some others, such as primary health and primary education, have been given to local governments. Third, major revenue sources such as commodity boards, company income tax, and all mining royalties and taxes have all been nationalized (see Elaigwu, 1992; Olowu, 1992; Oyovbaire, 1985).

In a sense, given the complete subordination of local governments to state/regional governments in the period up to 1976, any effort to strengthen local governments was bound to be at the expense of the state governments. During the civilian Second Republic, state governments struggled to reassert their lost powers—including the power to nominate local council members rather than choose them by popular election, the power to redraw local government boundaries, and the power to redefine local government responsibilities and their revenue sources. But subsequent military regimes, especially the present one, did everything to ensure that these powers were withdrawn from the states by asserting local government autonomy. The existence of democratically elected local governments was guaranteed by Section 7 of the 1979 constitution, and the 1989 constitution further enshrines local government autonomy by making local government responsibilities a permanent feature of the national constitution (see Table 8.1).

Having set the principle that local governments were entitled to a share of centrally collected revenues in 1978, the current administration increased the amount progressively from 10% (1981) to 15% (1990) and then 20% (1991). Furthermore, these monies were to be paid directly to local governments rather than through the state governments as in the past. Special directives were given to ensure that primary education and primary health care were transferred to local governments by federal and state agencies between 1986 and 1992. Local governments were also allowed to recruit their own staff, as the statewide Local Government Commissions, which carried out these responsibilities in the past, were abolished. In addition, a strong mayor system was introduced for all local governments throughout the federation, complete with separate legislatures.

Table 8.1 Functions of a Local Government (1989 Constitution)

1. *Exclusive*

The main functions of a local government shall be as follows:

a. formulation of economic planning and redevelopment schemes for the local government area;
b. collection of rates and insurance of radio and television licenses;
c. establishment and maintenance of cemeteries, burial grounds, and homes for the destitute or infirm;
d. licensing of bicycles, trucks (other than mechanically propelled trucks), canoes, wheelbarrows, and carts;
e. establishment, maintenance, and regulation of slaughterhouses, slaughter slabs, markets, motor parks, and public conveniences;
f. construction and maintenance of roads, streets, street lighting, drains, parks, gardens, open spaces, or such public facilities as may be prescribed from time to time by the military governor or house of assembly of a state;
g. naming of roads and streets and numbering of houses;
h. provision and maintenance of public conveniences, sewage and refuse disposal;
i. registration of all births, deaths, and marriages;
j. assessment of privately owned houses or tenements for the purpose of levying such rates as may be prescribed by the military governor or house of assembly of a state;
k. control and regulation of:
    ii. outdoor advertising and boarding;
    ii. movement and keeping of pets of all descriptions;
    iii. shops and kiosks;
    iv. restaurants, bakeries, and other places for sale of food to the public;
    v. laundries; and
l. licensing, regulation, and control of the sale of liquor.

2. *Participatory*

The functions of a local government shall include participation of such government in the government of a state as respects the following matters, namely:

a. the provision and maintenance of primary, adult, and vocational education;
b. the development of agriculture and natural resources, other than the exploitation of minerals;
c. the provision and maintenance of health services; and
d. such other functions as may be conferred upon a local government by the military governor or the house of assembly of a state.

Motivations for Local Government Revitalization

What are the interests the military has tried to promote through these reform programs? Four major reasons have been adduced for the reform of Nigerian local government. First, local government revitalization has been justified as part of the overall conception by the military elites that they constitute a "corrective" regime. That is, one of the ways in which they believe they can teach civilians about democratic governance, like the colonial officers before them, is through a revitalized local government system. This basic logic was played out during the colonial era as the colonial officers prepared the young nation for independence. Lord Creech Jones, the then secretary of state for the colonies, had argued the case for the institutionalization of an efficient and democratic system as a major strategy for grooming future political leaders and stimulating local economic initiative. In the same way, the first military rulers who prepared to return power to civilians after 13 years of military rule reasoned that local government reform was a crucial element in the political program of the federal military government (Nigeria, 1976, p. ii). This basic philosophy is also revisited in the second military transition program (1987-1992).

Military governments, like the colonial government before them, regard this reform of local governments as a political gesture demonstrating the seriousness of their commitment to civil rule. These reforms were expected to induce greater power sharing, reduce the prospects of tyranny in the postmilitary era, diffuse the pressure for the creation of additional states, and promote political integration through devolution and local administration of sensitive matters. In other words, local government revitalization is perceived as a crucial step in the redefinition of central-local relationships that was regarded as necessary in correcting a defective political structure.

A second motivation is economic. From the mid-1970s to the early 1980s, the country experienced an unexpected boom arising from the Arab oil crisis. Military elites saw the reorganization of local government as essential in ensuring that this unexpected windfall was channeled to address the problems of rural poverty, through the provision of social amenities. This became necessary in view of the general critique that government expenditures tended to be excessively centralized and were either urban oriented or skewed along ethnic lines. The federal military government articulated this concern as follows:

> These reforms would mean nothing if they did not include the certainty that as from now, every stratum of Nigerian society would benefit from the continued prosperity of this country, through the availability of amenities indeed, necessities, such as electricity, adequate water supply, improved transportation, health facilities and so on. (Nigeria, 1976, p. ii)

Third, when the tide turned in the 1980s and the country went into a deep economic recession, an even more important economic rationale emerged for revitalizing local governments. Local governments could be utilized to mobilize resources to pay for services that could no longer be funded from central government subsidies and transfers. Local governments became increasingly involved in the management of key services, such as basic health, education, rural roads, and water supply, to mention a few. In particular, the federal government mandated that services such as primary education and primary health care should be delegated to the local governments by the federal and state government agencies between 1986 and 1990. Each of these agencies resisted the move, but the federal government was insistent. For a country in which most of the private operators are outside the tax net, this had a lot of attraction for a government in serious fiscal crisis.

A fourth and final rationale for the reform is that it provided a formal channel of communication between the military government's administrative apparatus (at the federal and state levels) and the local people. This is a rationale that reinforces the first.

Achievements of the New Local Government System

Four major achievements can be claimed on behalf of this reform effort. First, it helped to conceptualize and articulate what constitutes local governments in the Nigerian milieu and their niche in the scheme of things. Up to that time, there was much confusion between local (field) administrations and local government. The huge transfer of resources from the national coffers to the new local government system made it possible for the local government share of total capital expenditure to increase from 2% in 1976 to an average of 6.6% between 1987 and 1989. This is not as high as it was in the 1950s and 1960s (which ranged from 11% to 15%), but, considering the huge increases in the size of public expenditure, the resources disposed by local governments have become considerable (see Tables 8.2 and 8.3). Both the fifth national development plan (1980-1985) and the first rolling plan (1990-

Table 8.2 Federal Transfers to Local Governments in Nigeria, 1976-1991

Fiscal Year	*Federal Transfers (in million N)*	*% of Federation Revenue*
1976	100.0	1.7
1977	250.0	4.2
1978	150.0	2.2
1979	300.0	2.5
1980	278.0	2.3
1981	1,085.0	9.0
1982	1,018.7	8.0
1983	996.8	9.1
1984	1,061.5	9.5
1985	1,327.5	8.5
1986	1,166.9	9.5
1987	2,117.8	8.4
1988	2,727.0	10.1
1989	3,399.0	10.0
1990	7,680.0	16.0
1991	10,765.0	19.1

SOURCE: Central Bank of Nigeria (1977-1991), Olowu (1990a).

1992) expected local governments to be responsible for more than 6% of total public sector expenditure in each respective period.

Arising from the increased fiscal activity of the local governments is a second achievement: Local expenditure programs went to the construction and maintenance of a range of basic economic and social services throughout the country. Local governments in the northern and southern portions of the country have about identical functions: markets, motor parks, local feeder roads, bridges, culverts, primary education, community health (including dispensaries, maternities, sanitary inspection, and provision of slaughterhouses/-slabs), information gathering, and public enlightenment. In the eastern states, local governments' activities also include the reconstruction of community halls, post offices, and other agencies.

An analysis of local governments' expenditure patterns indicates that Nigerian local governments devoted the larger share of their expenditures to recurrent items compared with capital expenditure items, the ratio being about 70:30. Whereas recurrent budgets are used primarily

Table 8.3 Share of Each Level of Government in Public Sector Expenditure for Selected Years, Nigeria 1955-1991 (in percentages)

Year	*Federal*	*State*	*Local*
1955	44.1	43.0	13.0
1965	44.5	44.7	10.0
1976	57.3	40.7	2.0
1985	64.8	29.8	5.4
1991	66.6	22.1	11.3

SOURCE: Adedeji and Rowland (1972), Central Bank of Nigeria (1977-1991).
NOTE: Rows do not add to 100 because of rounding.

to finance social services (especially elementary schools in the north, health services in the south), capital expenditures go for economic projects: markets, motor parks, roads, bridges, culverts, and the purchase of plants and machinery. Between the two types of services, economic and social, there has been greater investment in the latter because of the high share of the total expenditures devoted to recurrent items. The impact of this has been particularly noticeable in the northern states, where the child enrollment level for elementary education has increased from 19% (1976) to a figure that ranges from 28% (Bichi Local Government [LG], Kano state) to 69.6% (Barkin Ladi LG, Plateau state) by 1980. Administrative costs do not generally attract more than 20% of either recurrent or capital expenditures (Adamolekun, Olowu, & Laleye, 1988; Olowu, 1990c; Oyeyipo & Birkelbach, 1988).

The above findings are significant in at least three important respects. First, the revenue transfers through local governments ensured some form of equalization in the distribution of services in a large country such as Nigeria. They obliterated the sharp contrasts in local governments' effectiveness and performance in different regions of the country. Second, to the extent that local institutions devoted a substantial portion of their expenditures to basic social and economic services such as elementary education, basic health, markets, motor parks, roads, and some aspects of social welfare, they helped to tackle one of the major problems that has confronted the country: the sharp regional and geographic inequalities with respect to the availability of such services. Besides, these basic services are regarded as the cutting edge of local-level development (Filani, 1981; Stohr & Fraser-Taylor, 1981). In this sense, the transfers ensured that one of the major objectives of the reform of local government was achieved, at least in part. This was indeed one of the most important expectations of these revenue transfers: that access to basic infrastructures would be improved.

The Political Bureau, which was established in 1986 to advise Nigeria's military rulers on the appropriate structure of governance in postmilitary Nigeria, recommended the further strengthening of local governments through the transfer of responsibilities and resources currently controlled by the federal and state governments to the local governments. A number of responsibilities have been shifted downward from the federal/state governments to the local governments in the past few years; in addition, the proportion of the federation account going to local governments has increased from 10% to 20%. The resultant vertical sharing of the federation account since January 1992 is 48%, 24%, 20%, and 8% for federal, state, local governments, and special responsibilities, respectively.

A third achievement is increased citizen participation in the government system. In the first instance, the reform reaffirmed faith in local (self) governments rather than the trend toward local (field) administration in the country. This meant a strong emphasis on the representative and conciliar character of local governments. Second, the reform emphasized election rather than selection for filling local governments' membership positions. Even though modifications had been made in the 1950s and 1960s, the preponderant method for filling council positions up to this point continued to be selection. In addition, the huge monetary transfers from the federation account to relatively smaller local governments (compared with the larger, consolidated structures preceding the reform) provided opportunities for the emergence of a new crop of more sophisticated leaders at the local government level. Although this increased level of citizen participation is important in itself, the new local government system has attracted into the rural areas some urban dwellers who now form a new group of local politicians. The qualifications of the new cohort attracted to serve as political executives in local governments are quite high (lawyers, doctors, engineers) compared with the past penchant of having local governments dominated by councillors whose highest educational attainment was elementary school.

A fourth and final achievement is that the reformed local governments have experienced a number of other administrative changes that have boosted the prestige, status, and effectiveness of local governments' administrative and professional officers. These include improvement of personnel conditions through attracting more senior administrative and accounting personnel, consistent contributions to the training of their own staff and to a pension scheme instituted in 1979, as well as implementation of other staff welfare schemes such as housing and car loans.

One important outcome is that local governments today are playing critical roles in the implementation of a range of federal and state government programs in such areas as agriculture, electricity and water supply, and unemployment. Most of these programs are currently being coordinated at the federal level by a newly created rural development agency, the Directorate of Food, Roads and Rural Infrastructure. Local governments throughout Nigeria constitute the major implementation agencies of this body. Local governments are also involved in the implementation of the basic health scheme, the national identification scheme, population control, the expanded program of immunization, and so on. As a result, local governments currently enjoy strong support in Nigerian society.

The Low Degree of Local Government Institutionalization

In spite of these achievements, local governments remain as yet uninstitutionalized in Nigeria. Several factors are responsible, but a few can be noted. First, the rapidity of changes to the structure, responsibilities, powers, and size of local governments gives a sense of perpetual instability; this is one important factor that frightens many professionals who are being attracted to work in the local government system. A second factor is limited citizen control: Most of the central issues relating to the new local government structure are decided by either the state or the federal government. Indeed, between 1979 and 1987, local governments were managed either by management committees who were selected by the state governments or by civil servants seconded to the local governments as sole administrators.

Furthermore, local governments do not have much leeway in generating their own internal revenues, as the rates and bases of these taxes are controlled by the state governments. The budgets bear no resemblance to reality, as they are made to secure approval of state bureaucrats rather than for planning and budgeting for services. Similarly, local governments have powers to manage only junior-level personnel (levels 01-06); higher-level personnel are administered by Local Government Service Commissions, which see themselves more as part of the state apparatus than as independent agencies.[3]

Third, arising from a combination of the above problems and the huge transfers going to local governments in the absence of an incentive mechanism, local governments tend to be lax in collecting own revenue sources. In fact, local government own revenues have fallen steadily in *absolute* and *relative* amounts as federal transfers to local governments have increased. For instance, own revenue as a percentage of total revenues fell in relative and absolute terms between 1976 and 1990 as the volume of federal transfers to local governments rose from only N 100 million in 1976 to N 10.8 billion in 1991 (see Table 8.2). The fact that local governments derive virtually all their resources from outside has tended to reduce the accountability of these local governments to their citizenry.

If we define *local autonomy* in terms of local officials' ability to act without significant constraints from central government controls or from the constraints imposed by local economic and social groups (Clarke, 1989), Nigerian local governments possess neither form of autonomy. Important traditional rulers and other significant local notables can determine actions more effectively through their influence on state and federal government agencies than the local governments can. These problems, together with the penchant for sudden and major changes in local governments on a sustained basis, have led to a situation in which these official local governments are not regarded as reliable institutions for promoting socioeconomic development in Nigeria's rural communities, in spite of their achievements since 1976.

Community Development Associations: Institutions for Mediating Market and Community Relations

It is evident from the discussion so far that officially mandated local governments have been the subject of severe policy instability, even though the cumulative effect of the administrative and constitutional changes has been to empower them. They are also heavily dependent on state (financial) resources, management, and initiatives. In civilian periods, official local governments face less auspicious circumstances, and the struggle is always very severe between the contending political parties in their attempt to gain control of local governments.

In the face of this instability and high level of politicization, various communities, both urban and rural, have evolved institutional strategies for coping with their economic and social reality. Almost everywhere, nonofficial community development associations have been created to provide a number of public goods and services with or without government assistance. These indigenous community development initiatives represent local people's response to economic stimulus and are seen as having successfully "conserved those elements in the traditional political culture [that favored] development" and harnessed them to modern elements to develop their primary communities. These informal community governments play the role of de facto governments in many rural areas of Africa, as governments' presence is often confined to the major urban centers as a result of political, organizational, and communication problems. They provide the basic infrastructure (schools, health clinics, hospitals, public water, electricity, roads) needed to enable rural people to take advantage of economic opportunities (Akinbode, 1977; Awa, 1988; Jones, 1979; Ohachenu, 1990; Smock, 1971).

Variations in Community Development Associations

Recent research conducted in 10 of Nigeria's 21 states (prior to August 1991) describes CDA roles.[4] First, CDAs can be differentiated on the basis of their scales of operation (federated or national, regional or panethnic, ethnic, community, age/grade, clan or family), profession, gender, purpose, and other structural characteristics. For purposes of this study, research was limited to community- or town-based CDAs; we found that in most communities, the town/community government headed by a local chief was distinct from the CDA. CDAs are usually of two types: those sponsored by members of the community working and living outside that community, and those operated by members of the community living and working in it. CDAs can also be differentiated in terms of their structures and leadership types. Most have formal constitutions, often adopted from state model statutes, but others have only unwritten constitutions. In one CDA in Lagos state (Agonrin), all members are made to pledge to a local authority (*Zangbeto*) that also enforces sanctions against default. Some CDAs have voluntary membership, whereas several are compulsory for all sons and daughters of the community. Virtually all CDAs exact annual taxes or pledges on family heads that are usually directed toward the financing of specific projects. These "informal taxes" could be as high as N 2,000 a year per head, although the amount usually varies with income.[5] CDAs generally have democratically elected leaders, although some of the most successful have been led by one leader since their inception.

A second major finding is that "successful" cases of community development associations can be found in virtually all parts of Nigeria. This finding corrects the regional emphasis of the literature on local institutions in Nigeria. For instance, it has commonly been held that local governments are generally stronger in northern Nigeria because of the earlier success of the indirect rule system in that part of the country, and that CDAs are weak in that area. Strong and successful communal institutions such as CDAs are expected to exist where formal local governments are weakest—the southeast parts of Nigeria. These CDAs are reputed to have taken over the responsibilities of official but largely irrelevant local governments (Akinbode & Oye, 1988; Awa, 1988; Olowu, 1989). But the finding of widespread CDA success has important implications. One such implication is that tremendous opportunities exist for strengthening both local governments and community development associations in all parts of Nigeria by focusing on what makes for successful or not-so-successful performance. More important, there are tremendous opportunities for strengthening both through mutual learning and more purposive interaction.

A third major finding is that CDAs are playing a critical role in the provision of goods and services, both in the cities and in rural areas of Nigeria, with special reference to the latter. In fact, according to various reports, these local nonofficial community organizations constitute "the government" of several rural areas in Nigeria. The most important rationale for CDAs is that they make infrastructure available, especially in the rural areas, where government presence is usually least felt. The current economic crisis, by reducing the capability of central governments to finance infrastructure, has further heightened the need for such services and informal organizations that specialize in producing such services. CDAs have been particularly effective in the provision of physical and social infrastructure—roads, water, health care, elementary and adult education, library facilities, post offices, electricity extension, and street lighting. Although CDAs exist primarily to promote the construction and maintenance of public goods, a few of them engage in the production of private goods as well. The most common private engagement is the mobilization of credit; in recent years, as the economic crisis intensified, many have gone into the promotion of small-scale industrial enterprises.[6]

A fourth significant finding is that, comparatively, successful CDAs have greater impact than do successful local governments, in spite of the fact that more centrally collected financial resources increasingly have been made available to local governments in Nigeria since the nationwide reform of local government in 1976. Three reasons are adduced for this situation. First, local governments are not accountable to their citizens. Their primary and sole accountability until recently was to the state government, which decides their budgets, provides the most critical staff, and determines their priorities. Second, the political leadership of the local government is more often than not appointed by the state government rather then elected by the people. Indeed, since the local governments were reformed nationally in 1976, there have been only three elections, two of which were held within the past three years as part of the transition program back to civil rule.[7] State appointees have either been close confidants of the state governors or members of their political party, or, under military rulers, senior civil servants appointed as "sole administrators," that is, without a deliberative council or any separation between the legislative authority and the executive. And finally, as noted above, the large infusion of funds from the central (federal) government has made local governments lax about the generation of own resources and, indeed, the effective utilization of available resources. In contrast, the successful CDAs are completely accountable to their members and raise virtually all their financial resources internally. As a result, CDAs in many parts of the country were found to be performing as de facto local governments.

A fifth major finding relates to the problems confronting these CDAs. These problems are classifiable into exogenous and endogenous problems. The exogenous problems come from a combination of factors in the physical, economic, and political environments, and the endogenous problems include intracommunity rifts and lack of requisite professional personnel. Generally, CDAs were found to be strong in planning and implementing community projects but weak in terms of maintenance. Other weaknesses of CDAs are that there are few horizontal linkages among the associations and no strong formal vertical relationships between them and the formal political structures (especially local governments) (Enemuo, 1990).

In characterizing the factors that dispose CDAs to success or failure, the most significant factor seems to be leadership. The reports point to a dynamic relationship between effective, committed, and exemplary leadership on the one hand and faithful and dogged followership on the other. The quality of leadership tends to have a strong impact on the organization's structure and success. The quality of the local environment tends to have an inverse effect on the performance of CDAs: Poor rural environments tend to stimulate community development initiatives among members of a local community. Age was found to have the least relevance to local organizational performance, and linkages with other organizations have no determinative effect. CDAs are only tenuously linked to state and federal government agencies, but several successful CDAs are well linked informally to several public and private sector organizations through "sons and daughters of the soil." The upshot of all this is that it is the nature or content/quality of the linkage that matters, rather than the absence or presence of linkages. Individual representatives of the community who are by birth or marriage automatically members of the CDAs can utilize their influence to ensure that their communities benefit maximally from the programs of all levels of government.

Conclusion

The Nigerian situation demonstrates quite conclusively that local institutions are not only relevant to the development process, they have been undergoing changes both of a vertical and horizontal nature in order to ensure that they respond adequately to political and economic stimuli. For a Third World country that has experienced the sudden economic boom arising from oil price escalation and the depression arising from excessive dependence on a monoculture, the response of the local people to these varying opportunities is insightful. Similarly, the different roles of military leaders, who have virtually ruled the country since independence, and their civilian counterparts in redefining the relationship between central and local governments are equally captivating.

Awareness of these twin processes has increased among Nigeria's policy makers. In the past five years, for instance, the federal government has not only pressed for stronger local governments but formally recognized CDAs as institutions for mobilizing resources for economic development. As part of its mandate to collaborate with federal, state, local, and community development agencies, the DFRRI had identified a total of 114,000 CDAs by the end of 1990. Similarly, the federal government, in recognition of the role of CDAs in mobilizing resources for improving their communities, has created the Federal Board for Community Banks, which stimulates communities to establish rural banks by offering loans of up to half a million naira to match their own locally generated resources. The community bank scheme draws upon the principles of the indigenous credit associations and has, to date, been a resounding success in terms of paid-up capital, loan repayment by the membership, and growth of the banking units. Finally, it is perhaps needless to say that the Nigerian case is not peculiar. The relevance and role of local institutions in the politics and economies of African countries have increased with the economic crisis.

Notes

1. Because of the political advantages associated with numbers, population censuses in postindependence Nigeria have been dogged by controversy. The 1963 census was rejected, but when the 1973 census count was canceled, the 1963 census was accepted as official until another count in 1991 suggested that Nigerian population size has been overestimated (88.5 million). This latest count has also been hotly contested by many states; some have sought redress in the Population Tribunal.

2. For an overview and robust discussion of the 1976 local government reform, see Gboyega (1983), Adamolekun (1984), and Olowu (1990a).

3. The Local Government Service Commissions were abolished in January 1992 precisely because they were thought to compromise the principle of local government autonomy. After local government employees brought pressure to bear on the federal government, however, they were resuscitated.

4. Some of the general information in this section comes from Olowu, Ayo, and Akande (1991) and Ayo, Hubbell, Olowu, Ostrom, and West (1992). Funding for the research was obtained from the Ford Foundation, West Africa Office.

5. In 1990, U.S. $1 = N 10.00; currently, U.S. $1 = N 18.00.

6. Three successful CDAs illustrate these variations. The Abwa-Mbagben Development Association in Benue State had in 1989 an annual income well over N 1 million. Among other things, it has established three secondary schools and rehabilitated a number of primary schools. It built a 50-bed health center, complete with an operating theater, and pays for the services of one of the two doctors; the other doctor is provided by the state government, as is an ambulance. Before this time, no health center existed in the area. It has also built several roads and bridges to make travel by car in the area easier.

A second example is a federation of several CDAs run by women in Ondo State, the Community Women Association of Nigeria (COWAN). COWAN began in Ondo in 1982 as an apex voluntary development organization with the sole aim of organizing women into cooperatives all over the federation. By 1990, it had state chapters in 10 of the 21 states in the country and a national membership of some 15,000 people (mainly women). In Ondo State it established more than 450 women-only cooperative societies. Its activities include operating a revolving loan scheme started in 1984 on the basis of the traditional *esusu,* or savings club. The success of this scheme led the Union Bank in 1989 to grant a loan of N 2.7 million to COWAN chapters in Ondo, Bendel, and Rivers States. Other activities include an integrated health family planning education and services project in collaboration with the United Nations Center for Economic Development and Population

Activities, fabrication of simple gari-processing technology, training of trainers, and production of training materials.

A third example is the Igboho CDA of Oyo State. Its activities since its inception in 1954 have included the provision of a dispensary, police station, postal agency, town hall, central market (with 20 open stalls), motor park, secondary schools, a laboratory for secondary schools, a general post office and staff quarters, purchase of a 504 Peugeot van for facilitating mail delivery, town bus transportation, and provision of scholarships to deserving indigents.

7. Three local government elections have been conducted since 1976: in 1987, 1989, and 1991. The level of voter turnout at the 1976 election (about 30%) was not much below that for the national presidential election in 1979 (36%), whereas the 1987 voter turnout was higher than in 1976 (about 50%). This was in spite of the special circumstances in which the elections were organized: They followed a long period of military rule without any elections and the suspension of political parties. Subsequent elections have shown similar patterns of voter turnout.

References

Adamolekun, L. (1984). "The idea of local government as a third tier of government" revisited: Achievements, problems and prospects. *Quarterly Journal of Administration, 18,* 92-112.

Adamolekun, L., Olowu, D., & Laleye, M. (Eds.). (1988). *Local government in West Africa since independence.* Lagos: Lagos University Press.

Adedeji, A., & Rowland, L. (Eds.). (1972). *Local government finance in Nigeria.* Ile-Ife: University of Ife.

Ake, C. (1987). *Sustaining development of the indigenous.* Washington, DC: World Bank.

Akinbode, I. A. (1977). Participation in self-help projects among rural inhabitants. *Quarterly Journal of Administration, 11,* 285-298.

Akinbode, I. A., & Oye, M. (1988). The role of voluntary association in the development of local government areas. In L. Adamolekun, D. Olowu, & M. Laleye (Eds.), *Local government in West Africa since independence* (pp. 239-264). Lagos: Lagos University Press.

Awa, E. (1988). The transformation of rural society in the eastern states of Nigeria, 1970-76. In L. Adamolekun, D. Olowu, & M. Laleye (Eds.), *Local government in West Africa since independence* (pp. 221-238). Lagos: Lagos University Press.

Ayo, S. B., Hubbell, K., Olowu, D., Ostrom, E., & West, T. (1992). *The experience in Nigeria with decentralization approaches to local delivery of primary education and health services.* Burlington, IN: Associates in Rural Development.

Central Bank of Nigeria. (1977-1991). *Statement of account.* Lagos: Author.

Clarke, S. E. (1989). Urban innovation and autonomy: Cross-national analyses of policy change. In S. E. Clarke (Ed.), *Urban innovation and autonomy: Political implications of policy change* (pp. 21-30). Newbury Park, CA: Sage.

Elaigwu, J. I. (1992). Nigerian federalism: Experiences under civilian and military regimes. In D. Olowu (Ed.), *Federal administration in Nigeria.* Ibadan: Evans.

Enemuo, F. C. (1990). *Communal organisations, the state and rural development: The political economy of community self-help efforts in Anambra State, Nigeria.* Unpublished doctoral dissertation, University of Lagos.

Esman, M. J., & Uphoff, N. T. (1984). *Local organization: Intermediaries in rural development.* Ithaca, NY: Cornell University Press.

Filani, M. O. (1981). Nigeria: Need to modify top-down development. In W. B. Stohr & D. R. Fraser-Taylor (Eds.), *Development from below or above?* (pp. 283-304). Chichester: John Wiley.

Gboyega, A. (1983). Local government reform in Nigeria. In P. Mawhood (Ed.), *Local government in the Third World: The experience of tropical Africa.* Chichester: John Wiley.

Guyer, J. (1991). *Representation without taxation: An essay on democracy in rural Nigeria* (Working Papers in African Studies No. 152). Boston: Boston University, African Studies Center.

Hicks, U. (1961). *Development from below: Local government and finance in developing countries of the Commonwealth.* Oxford: Clarendon.

Hyden, G. (1986). Urban growth and rural development. In G. M. Carter & P. O'Meara (Eds.), *African independence: The first twenty-five years* (pp. 188-217). Bloomington: Indiana University Press.

Jones, G. I. (1979). Changing leadership in eastern Nigeria: Before, during and after the colonial period. In N. A. Shack & P. S. Cohen (Eds.), *Politics in leadership: A comparative perspective* (pp. 44-64). Oxford: Clarendon.

Nigeria, Federal Republic. (1976). *Guidelines for local government reform.* Kaduna: Government Printer.

Nigeria, Federal Republic. (1986). *Report of the Political Bureau.* Lagos: Government Printer.

Ohachenu, U. E. (1990). *Survey of the role of non-governmental agencies in rural development: Anambra State.* Lagos: Directorate of Food, Roads and Rural Infrastructure.

Olowu, D. (1988). Local government and grassroot democracy in Nigeria's Third Republic. *Nigerian Journal of Public Policy, 3,* 78-97.

Olowu, D. (1989). Local institutions and development: The African experience. *Canadian Journal of African Studies, 22,* 201-231.

Olowu, D. (1990a). Achievements and problems of federal and state transfers to local governments in Nigeria since independence. In L. Adamolekun, R. Robert, & M. Laleye (Eds.), *Decentralization policies and socio-economic development in sub-Saharan Africa* (pp. 116-156). Washington, DC: Economic Development Institute.

Olowu, D. (1990b). Local government. *Quarterly Journal of Administration, 24,* 179-189.

Olowu, D. (1990c). *The Nigerian conception of local level development.* Ibadan: Nigerian Institute of Social and Economic Research.

Olowu, D. (1992). *Federal administration in Nigeria: Experiences under military and civilian governments.* Ibadan: Evans.

Olowu, D., Ayo, D., & Akande, B. (1991). *Local institutions and national development in Nigeria.* Ile-Ife: Obafemi Awolowo University Press.

Olowu, D., & Smoke, P. (1992). Determinants of success in African local governments: An overview. *Public Administration and Development, 12,* 1-17.

Oyeyipo, E., & Birkelbach, A. (1988). Changes in the functions performed by local governments in the northern states of Nigeria. In L. Adamolekun, D. Olowu, & M.

Laleye (Eds.), *Local government in West Africa since independence* (pp. 133-154). Lagos: Lagos University Press.

Oyovbaire, S. E. (1985). *Federalism In Nigeria: A study in the development of the Nigerian state.* London: Macmillan.

Smock, A. (1971). *Ibo politics: The role of ethnic unions in eastern Nigeria.* Cambridge, MA: Harvard University Press.

Stohr, W. B., & Fraser-Taylor, D. R. (Eds.). (1981). *Development from below or above?* Chichester: John Wiley.

Wunsch, J. S., & Olowu, D. (Eds.). (1990). *The failure of the centralized state.* Boulder, CO: Westview.

9

Political Restructuring and the Development Process in Kenya

PETER M. NGAU

The countries of sub-Saharan Africa are currently going through a period of profound change in their economic and political institutions. Since the 1980s, these countries have been going through an economic recession of far greater magnitude than any faced since the independence years of the 1960s. According to the World Bank's (1990) *World Development Report,* sub-Saharan Africa is the only major region of the world where the incidence of poverty increased over the past decade. Indeed, the period of the 1980s was a lost decade for sub-Saharan Africa in economic terms. Production levels have fallen in agriculture and industry, and basic services and infrastructure have deteriorated. In the past five years, sub-Saharan Africa has also faced such political turbulence as never witnessed since the days of independence struggles. Rapid political changes are taking place from one-party and military dictatorships to greater democracy, respect and concern for human rights, and people's greater participation in all matters relating to their lives. At the local level, there has emerged greater expression of political activism and an explosion of associational activity demanding democratization, accountability, and openness in public affairs.

The economic and political changes occurring in sub-Saharan Africa are affecting fundamentally the way the development process is viewed

AUTHOR'S NOTE: I gratefully acknowledge the assistance of Dr. Willy Mutunga with information concerning recent political pressure groups and legal organizations in Kenya.

and conducted both at the center and at the local level. New strategies for economic development have begun to emerge in the face of global economic changes and the increasing marginalization of the continent. As well, new forms of relationships among central government, local government, and civil society are being defined. This chapter illustrates the extent of these changes as they are unfolding in Kenya. First, I seek to place in historical context the development process in Kenya since independence, including past negotiations of changes in the relations among the central government, local government, and the role of development planning. I will then explore the range of recent experimentation and innovation in economic development strategies and political organization occurring in the country.

Among the new economic development strategies are the Rural Trade and Production Centres (RTPC) program and the Small-Scale Enterprise program. Related programs for institutional restructuring include the Local Authority Development Programme (LADP) and a renewed effort to strengthen the district planning process. In the political sphere, there is the emergence of local policy activism and political agency in support of the environment as well as economic, social, and gender interests, respect for human rights, and greater participation in the development process. Private sector organizations have also emerged to safeguard the interests of business in various ways.

Overview of the Development Process in Kenya: 1963-1992

At independence in 1963, Kenya set out to establish the institutional machinery for countrywide development planning and administration (Kenya, 1965a). Independence gave expression to high expectations from indigenous Kenyans demanding a wide range of social and economic services denied to them during the colonial period. Like many newly independent countries, Kenya embraced the popular doctrine of comprehensive national planning for promoting rapid economic development.

In the ensuing years, Kenya managed to establish an elaborate machinery for development planning; during the first decade after independence, the country achieved very rapid economic growth as well as an impressive social and economic infrastructure network. Figure 9.1 depicts the rapid growth of Kenya's economy between 1962 and 1970, at a time when its population growth was relatively low. The rapid economic growth in the 1960s was largely the result of the expansion of smallholder agricultural production (Chege & Kimura, 1986; Heyer, Ireri, & Moris, 1971; Ngau, 1989). From the early 1970s, however, the country faced a series of economic crises (sharp oil increases, falling prices of agricultural commodities) and a rapidly growing population. As a result, the rate of economic growth began to decline, as shown also in Figure 9.1.

Beginning in the mid-1970s, efforts were made to respond to the economic problems facing Kenya. The government embarked on a program of extending development planning to the districts. It also set out to implement a spatial systems program aimed at providing a hierarchy of service and growth centers linked by a network of roads and communication technologies. However, during the 1970s Kenya had only modest success in implementing both the district planning and the spatial systems program because of the deepening economic recession, stagnation in agricultural and industrial production, and continuing population growth (Hunt, 1984; Livingstone, 1981). In 1982, there was an attempt to overthrow the government, followed by severe drought and famine in 1984.

In a report titled *Economic Management for Renewed Growth*, the government set forth a new framework of economic restructuring to revitalize the economy (Kenya, 1986). The country's economic performance improved in the 1985-1988 period because of improved commodity prices and high rainfall, but began to decline again in 1989. From 1989 on, the growth rates of real earnings and wage employment (which had been stagnant since 1979) began to decline rapidly, as shown in Table 9.1.

Most recently, in addition to these economic dilemmas, demands for multiparty politics are setting in motion an unprecedented momentum for political restructuring. There is an explosion of greater local policy activism on environmental and social issues, a search for new approaches to economic development, and concerns for local entrepreneurial roles.

The Origins of Development Planning and Politics in Kenya: 1963-1973

During the colonial period, planning consisted of ad hoc programs concerned mainly with capital projects. Comprehensive development

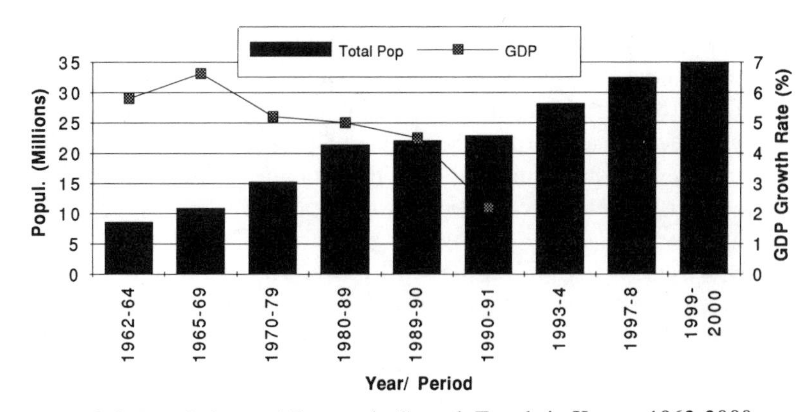

Figure 9.1. Population and Economic Growth Trends in Kenya, 1962-2000
SOURCE: Kenya, *Economic Survey* (various years); Central Bureau of Statistics, *Population Reports.*

planning emerged in Kenya and took shape soon after independence; it was embraced as a means of achieving rapid economic development and social change. Planning would guide the allocation of scarce resources—land, skilled human labor, capital, and foreign aid (Kenya, 1965a).

A major issue at the time of independence concerned how to share political power and bring about national unity (necessary for rapid economic development) in a multiethnic state. There were two contending proposals: One called for a federalist government, whereas the other favored a strong unitary government. There were also divisions concerning what economic system Kenya should pursue, between those who favored a socialist/communist path and those who preferred market capitalism.[1] For a brief period between 1963 and 1965, the country experimented with a federalist political system called *majimbo.* The *majimbo* constitution was a result of a coalition government between the Kenya African National Union (KANU) and the Kenya African Democratic Union (KADU); it sought to provide for a central government and strong regional governments called assemblies. The assemblies enjoyed strong political autonomy and had authority over local government agencies (at the districts) that were responsible for providing basic services such as health, education, and roads.

After 1964, the KADU party dissolved itself to join KANU in a unitary government. Thereafter, the KANU-led government moved rapidly to scrap the *majimbo* constitution and to build a strong unitary government. The government also decided to pursue a program of

Table 9.1 Recent Trends in GDP, Real Earnings, Cost of Living, and Wage Employment in Kenya, 1986-1991

Year	*GDP Growth Rate (%)*	*Growth in Real Earnings (%)*	*Growth in Employment (%)*	*Average CPI (1975 = 100)*	*% Change in CPI*
1986	5.50	—[a]	—[a]	355.00	5.60
1987	4.80	—[a]	—[a]	380.00	7.10
1988	5.10	6.90	4.30	420.00	10.70
1989	5.00	−3.60	1.70	464.00	10.60
1990	4.30	−5.80	3.00	523.00	12.60
1991	2.20	−8.30	2.30	590.00	19.60

SOURCE: Kenya (1992a).
a. Not available.

economic development based on market capitalism, which it conveniently labeled "African socialism" (Kenya, 1965a). Thus the 1963-1965 period was characterized by sharp political divisions and interethnic conflicts over power and resource sharing between the regional assemblies. The first development plan produced for the period 1964-1970 proved difficult to implement (Gertzel, Goldschmidt, & Rothchild, 1969).

The decision to establish a centralist government under a unitary political system and to promote private enterprise has had considerable implications for the evolution of development planning, the balance of power between central and local governments, and economic development in Kenya. Starting in 1965, steps were taken to institutionalize a strong central planning and administrative framework extending down to provinces, districts, divisions, and locations (Kenya, 1965a). The main activity of national development planning and coordination was placed under a ministry then named the Ministry of Economic Planning and Development (MEPD).[2] The MEPD was responsible for supervision, direction, and control of economic policy and overall development planning in the country. By 1970, the MEPD was strong; it consisted of four principal divisions: Sectorial Planning, Macro Planning, Rural Planning, and the Central Bureau of Statistics. These divisions now form the pillars of development planning in Kenya.

During this period, local government authorities in Kenya gradually lost their power and local autonomy.[3] Three main factors contributed to their decline. First, local government agencies were unpopular at the time of independence, especially among the nationalist leaders, because they were staffed mainly by appointed loyalists during the colonial period and were used to levy taxes and suppress the struggle for freedom in the local areas. Second, after the fall of the *majimbo* constitution, the unitary government wished to assert itself. In 1965, the Hardacre Local Authorities Commission transferred many of the powers exercised by local authorities to the Ministry of Local Government (Kenya, 1965b). The commission gave this ministry power to dissolve local authorities for periods of two years and to nominate qualified people to the authorities. Third, local authorities began experiencing financial problems because they had lost their revenue base to the central government and so were increasingly dependent on the central government. By 1969, the fiscal situation of the local authorities became extremely precarious, and the central government assumed responsibility over three basic services previously provided locally: primary education, health, and secondary road maintenance.[4] In essence, the 1965 and 1969 reforms of local authorities in Kenya had the effect of reinforcing the administrative and technical jurisdiction of the central government at the expense of local government authorities.

Kenya's record economic growth during the first development decade in many ways appeared to validate its choice of a market/free enterprise model of economic development. The government concentrated resources in the high-potential areas and extended agricultural settlement of Africans into the former "white highlands," thereby boosting smallholder cash crop production tremendously. The increased yield and revenue from smallholder producers were mainly responsible for the accelerated economic growth (a 7% GDP growth rate) of the country in the 1960s and early 1970s (Heyer et al., 1971; Ngau, 1989). President Jomo Kenyatta's call for *harambee* also played a big role in inspiring national- and local-level enthusiasm for development.[5]

Although the government was successful in establishing a strong Ministry of Economic Planning and Development at the center, it had less success in extending planning activity to lower administrative units during the first decade. A two-tier hierarchical committee system was proposed as early as 1965 for purposes of coordinating and extending planning to lower administrative units in the country. The first tier included Provincial Development Committees (PDCs) and District Development Committees (DDCs) staffed by senior officials of the operating ministries at the provincial and district levels. The second tier included Provincial Development Advisory Committees (PDACs) and District Development Advisory Committees (DDACs) consisting of citizen volunteers and representatives of local groups to be involved in planning and mobilizing for development at the local areas.[6] Both the Provincial and District Development Committees remained weak, because the country did not have enough officials skilled in development administration and planning to staff these committees fully. A report in the late 1960s described both PDC and DDC meetings as infrequent, poorly attended, and unproductive (see Heyer et al., 1971). The DDACs were never constituted. As a result, throughout the 1960s, development planning in Kenya remained the prerogative of central government.

Attempts to Extend Development Planning to the Districts: 1972-1985

Beginning in the early 1970s, Kenya faced its first major economic crisis, the extent of which was captured in surveys by a mission from the International Labor Organization (ILO) and by a government commission called the Ndegwa Commission of Inquiry (see International Labor Organization, 1972; Kenya, 1971). The ILO report praised the performance of the Kenyan economy in the first decade, but also underlined its weakness with regard to employment creation and poverty alleviation. According to the ILO report, the performance of the Kenyan economy in the first decade was quite remarkable in terms of growth, but disappointing with respect to the growth of employment and its impact on the poverty of the lowest income groups. The Ndegwa Commission of Inquiry (1970-1971) focused on the role of government in development administration. Its report criticized development planning and administration in Kenya for failing to integrate development activities at the district level in the overall national development planning process and for the inefficiency of the Kenyan bureaucracy at the top.

In response to these reports, the Kenyan government began working on an array of innovative measures designed to address the socioeconomic and institutional challenges that faced the country. The Second Development Plan (1970-1974) announced the implementation, on an experimental basis, of a project called the Special Rural Development Program (SRDP). The SRDP called for initiation of integrated development projects in designated local areas in the country; the plan, the outcome of a conference in Kericho in 1966, detailed how spatial systems planning (the service and growth center strategy) would be implemented in the country. The implementation of Phase I of SRDP did not start until 1971; at the end of 1972, the Kenyan government made the decision not to move into Phase II but instead to establish district-based planning throughout the country.[7]

Three factors prompted the government to attempt to extend development planning to the district level. First, growing regional inequalities, vividly revealed by the ILO report, required more decentralized development planning to ensure effective implementation of projects at the local level. Severe regional inequalities posed a threat to political stability and national unity, and the government wished to assure the underdeveloped areas that they would also enjoy the fruits of independence. Second, as the Ndegwa Commission pointed out, development administration was becoming so complex that it could no longer be well managed entirely from the center. Third, it was becoming clear to economic planners that further integration of the country into the market economy required opening up the country by extending infrastructure and modern institutions nationwide.

In order to establish district-based planning, it was necessary to effect changes in the relations between the central government and the local government, in the processes of development decision making, and in the allocation and distribution of development resources. In the Kenyan context, these changes would revolve around a continuing tension between those who favor retention and/or strengthening of local community planning (and local government) and those favoring technocratic and bureaucratic institutions to plan and implement local development. The government choice was to implement a framework of district-based planning proposed by the Ndegwa Commission. It emphasized formal technical planning and provided for deconcentration of central government functions by moving decisions formerly made at the center downstream to subordinate units of the same ministry. However, the implementation of both district planning and spatial systems planning programs did not start until the mid-1970s because of economic constraints brought about especially by sharp increases in world oil prices in 1973.

During the Third and Fourth Development Plan periods (1974-1978 and 1979-1983), the Kenyan government was able to implement some of the ideas developed in the early 1970s concerning district planning and the service and growth center strategy. Several steps were taken to implement planning activities at the districts, including training and posting of district development officers (DDOs) to each of the 41 districts in the country, preparation of development plans for each district, and initiation of funding sources in support of district planning. Two main funding sources were established during this period: the Rural Development Fund (RDF) and the European Economic Commission (EEC) fund for micro-level projects.[8] Work started on preparation of district plans by the reorganized DDCs using guidelines developed by a new unit in the MEPD called the Rural Planning Division. Arrangements were also made for DDOs to be trained in project preparation and development planning, with assistance from the Institute for Development Studies at the University of Nairobi and the Kenya Institute for Administration. In 1976, the government also contracted with Harvard Institute for International Development (HIID) to provide technical assistance to Kenya, focusing on district-level planning, beginning with the preparation of district development plans for the 1979-1983 national plan period.

By the end of 1982, Kenya had made some progress in extending development planning to the districts. However, district planning continued to face many problems and remained unfulfilled in many aspects (Cohen & Hook, 1986; Delp, 1980). For example, translating district initiatives into real programs and channeling them into the operational ministry budgets and schedules remained a persistent problem (Cohen & Hook, 1986). And the mechanisms for implementing development programs still remained highly vertical within individual ministries, rather than being horizontally coordinated. District development plan making suffered from a lack of human resources at the district level; hence plans were written largely at the center (Delp, 1980).

Assessments from outside the government pointed to more fundamental weaknesses in the framework of district planning that was emerging in Kenya. After examining the process of budget estimate preparation at the districts and in the Ministries of Agriculture and Livestock Development, Chege and Kimura (1982) observed that the preparation of budget estimates was given very low priority at the district level. At the ministry level, allocations to districts tended to be quite arbitrary and were largely based on bargaining with district officers and political patronage. Oyugi (1981) and Makokha (1985) argue that district planning in Kenya tends to minimize the involvement of local people in the initiation, planning, and implementation of local development and that participation of local people is allowed only in modes acceptable to the state. Evans (1988) has also pointed out that lack of coordination between ministries was a major constraint that contributed to patchy implementation of the growth center and service program in the 1970s and 1980s. The program was "housed" in the Ministry of Physical Planning, which was allocated only managerial and operational funds (versus development funds). The ministry was also in a weak position to implement the program, and the other ministries had their own sets of priorities.

In 1983, the Kenyan government announced a program called District Focus for Rural Development (DFRD) in an attempt to revitalize district-based planning. DFRD was intended to delegate the operational responsibility for rural development to the districts and to encourage local development efforts and initiative. HIID was assigned to develop the necessary planning and financial systems by formulating and issuing guidelines and criteria to be followed in budget estimate preparation and district development plan making.

By 1986, a lot had been done to overcome the constraints facing district planning in the areas of human resources training and posting, and institutionalization of a mechanism for incorporating district project proposals into ministerial forward budgets to the treasury. Ministry estimates are now disaggregated by district and presented in an annual printed volume called *District Allocations.* Efforts to improve the technical capacities of government officials at the district level in formal planning were also successful. However, other fundamental goals of district planning remained unfulfilled. Among these were the needs to raise the effectiveness of project implementation at the local level, to allow for greater participation of local communities in the development process, and to promote development initiatives that address local development needs and priorities as well as national objectives.

Essentially, district planning in Kenya has succeeded in deconcentrating central government functions and empowering government officials both at the center and at lower levels.[9] Coordination, however, continues to be a problem, as the operation of government ministries and agencies remains highly vertical and each ministry tends to have its own priorities. The continued use of top-down planning and implementation methods has the effect of minimizing the involvement of local people in the initiation, planning, and implementation of local development programs. The DDCs, Division DCs, and Locational Development Committees are composed mainly of government technical and administrative officers. Local public meetings (called *barazas*) are also controlled by the government administrators (*chiefs*), who have the authority to decide who may attend and what has to be deliberated in the meetings.[10] Over the years, the role of political patronage has also expanded extensively, distorting the development planning process and grassroots participatory activities such as the *harambee* (self-help) movement (Ngau, 1987).

The DFRD tried to provide a mechanism for project identification and selection that would ensure that the majority of development projects from the districts (forwarded to the ministries for funding) originate in local areas (subdistrict level) to reflect local participation and relevance of local needs. However, it appears that, in practice, very few projects forwarded to the ministries through the district planning process actually originate in the local areas. Furthermore, the majority of development projects in the districts that are finally funded through the ministries are rarely oriented to addressing local development needs and priorities. Analysis of development projects forwarded from districts to the ministries during the 1985-1986 fiscal year shows that the majority of the projects still originated from department heads at the district headquarters and from provincial headquarters. In the DFRD, projects originated from three main sources: (a) locations and sublocations (local communities) through the Division DC, (b) department heads in the district, and (c) the DDC, submitted by a district agency or higher up from the provincial administration. Table 9.2 shows an analysis of Nyeri and Machakos districts, where the share of nonlocal projects was greater than 85%.[11] It is clear from the data that the level of involvement of local people in the identification and selection of development projects is insignificant.

There is also concern over the extent to which development projects in the district planning process address local needs and priorities. Three types of small projects are recognized under the DFRD strategy: (a) projects that are directly productive (e.g., cattle dips, poultry keeping, irrigation, and storage structures), (b) economic infrastructure projects (e.g., access roads and transportation, bridges, soil conservation, water supply, seedling nurseries, crop processing), and (c) social infrastructure projects (e.g., office buildings, office furniture, residential housing, education, and health). Analysis of projects forwarded from districts (1985-1986) again shows that projects of social infrastructure type were by far dominant in all the priority categories. It also reveals that most of the social infrastructure projects originated from government civil servants and benefited them more directly, as they were intended to improve the standard of living and working conditions of the civil

Table 9.2 District Project Priority Ranking by Project Origin, 1985-1986 Fiscal Year (in percentages)

	Project Source/Origin		
Priority	*Local Area (Division DC)*	*District Department Heads*	*DDC/ Provincial Heads*
Nyeri District			
High	16.7	74.0	9.4
Medium	13.5	69.2	17.3
Low	10.2	62.7	27.1
All	14.0	69.6	16.4
Machakos District			
High	17.5	63.5	19.0
Medium	8.2	72.1	19.7
Low	16.9	52.5	30.5
All	14.2	62.8	23.0

SOURCE: Compiled from the annual annexes (1985-1986) to District Development Plans for 1984-1988.

servants at the districts. Table 9.3 shows an analysis of Nyeri and Machakos districts with respect to project type. Projects originating in local communities were found to be more oriented to direct production and economic development, whereas those from district headquarters tended to be social-oriented projects, often very large, costly, and symbolic. It was also found that many projects were being moved up as high priority, far more than the funds were available, because prioritization was weak.[12]

Recent Changes: Experimentation, Political Restructuring, and Local Policy Activism

The momentum for the current economic and political changes taking place in Kenya began in early 1980s. At that time the country began experiencing the severe effects of structural adjustment programs (SAPs) introduced in Kenya since 1980 under the recommendation of the World Bank and the International Monetary Fund.[13] There was also a major

Table 9.3 District Project Priority Ranking by Project Type, 1985-1986 Fiscal Year (in percentages)

	Project Types			
Priority	*Directly Productive*	*Economic Infrastructure*	*Social Infrastructure*	*All*
Nyeri District				
High	14.4	25.0	60.6	49.3
Medium	14.0	14.0	72.0	23.7
Low	17.5	21.1	61.4	27.0
All	15.2	21.3	63.5	100.0
Machakos District				
High	18.8	20.3	60.9	34.4
Medium	19.4	9.7	71.0	33.3
Low	18.3	5.0	76.7	32.3
All	18.8	11.8	69.4	100.0

SOURCE: Compiled from the annual annexes (1985-1986) to District Development Plans for 1984-1988.

drought and sharp decline in the prices of major export agricultural commodities, which brought about a famine and economic recession. These hardships inevitably contributed to social discontent and political instability. Political tension intensified in the wake of government purges of political activists following the attempted coup of 1982.

In 1986, the government came up with a new set of economic proposals, contained in a major policy paper titled *Economic Management for Renewed Growth* (Kenya, 1986). New changes to enhance the District Focus for Rural Development strategy were also outlined in a revised document (Kenya, 1984). As a result, a number of new economic development strategies and institutional changes have emerged since 1986. Among these is a strategy for managing rural and urban development called Rural-Urban Balance. There are three components to the strategy: the Rural Trade and Production Centres program, the Local Authority Development Programme, and the Small-Scale Enterprise program. Other important policy initiatives recently introduced in Kenya are those for promoting greater roles for the private sector and business entrepreneurship through enterprise promotion zones, a parastatal reform program, and trade liberalization.

Through the revised District Focus strategy, the government has also set forth an array of policy and institutional measures designed to create an enabling institutional environment for district planning process that is supportive of participatory and sustainable development. In an effort to strengthen local government authorities and broaden their tax bases, several fiscal and revenue-enhancement programs are being tried. They include the introduction of local service charges, cost sharing, and revenue sharing between central government and local government authorities.

The RTPC Program

Kenya is experimenting with a new project, the Rural Trade and Production Centres program, aimed at accelerating economic growth, raising per capita income, and creating labor employment in both rural and urban areas.[14] The RTPC program is designed to use small urban centers to increase the agricultural productivity of their hinterlands and increase nonfarm employment opportunities in the small towns. It is recognized that in order to raise levels of production and productivity in agriculture, there has to be widespread adoption of modern production methods, which implies greater use of agricultural inputs and services, improved systems of collection and marketing of agricultural outputs, and a wider range of activities for processing agricultural produce. Further, growing incomes will result in increased demand for higher-order goods and services. It is thus necessary to increase the number and expand the size of market centers in order to enable them to serve as suppliers of inputs, markets for outputs, agroprocessing centers, and providers of consumer-demanded goods and services.

The RTPC program is an example of a policy that is based on market-based planning (Gaile, 1992). At first glance, market-based and planning policy may seem to be a paradox. Much of regional and urban development planning policy was based on the assumption that the market had failed to provide for regional equity. Given this assumption, it is understandable that the market was not used to remedy regional problems. Instead, non-market-based subsidy-oriented policies with heavy emphasis on infrastructure predominate in regional development planning. Market-based regional development policy prioritizes government investment based on the ability of that investment to facilitate, expand, and improve the functioning of the market system (Gaile, 1992).

The RTPC strategy contains several components of market-based policy. One important aspect is market expansion. Kenyan farmers who have poor access to market often grow crops for home consumption only. It is argued that these subsistence farmers could become more productive if their access to the market were improved and they could choose to switch part of their production from production for home consumption to production for sale in the market, thus upgrading the value of production overall. Another market-based aspect of the RTPC policy is the removal of market imperfections. Relevant programs are those concerned with removing restrictions to free entry and competition and providing needed information, skills, and credit to producers and entrepreneurs. A third aspect of the RTPC policy is provision of productive infrastructure and improvement of rural-urban linkages. Lack of infrastructure is a major constraint retarding trade, private capital flow, and diffusion and exchange of innovations.

The implementation of the RTPC program began in 1987. Donor interest in the RTPC program was high, and the first-phase funding was expeditiously secured from government and donor sources. The first phase is now complete; eight RTPCs have been developed at a cost of U.S. $5 million. The locations for the second phase have been chosen, and research has been conducted to improve the criteria and guidelines used in identifying and selecting RTPC candidate centers and types of investment (Ngau & Gaile, 1992).

The Small-Scale Enterprise Program

The major challenge faced by the Kenyan government today is unemployment. Modern wage employment accounts for 15% of the labor force in Kenya. In 1988, an estimated 1.7 million persons were engaged in the modern sector in the whole country. Out of this, 35,000 were engaged in the informal sector and small-scale enterprises. More important, the small-scale enterprise sector has recorded a significant growth of employment at a high rate of 10% per year since 1985, compared with an employment growth rate in the modern (large) sector of less than 3.0% (Kenya, 1992a). Studies show that the informal sector accounts for a far greater percentage of urban employment in Kenya than is captured in official estimates (Ondiege & Aleke-Dondo, 1991).

A program for promoting the growth of small-scale enterprises (SSEs) therefore has been initiated recently in Kenya. The enterprises targeted by the program are primarily small-scale manufacturing and commercial activities, the bulk of which are in the informal sector.[15] Measures being taken to promote SSEs aim at creating an enabling environment in which SSEs can thrive; they consist of (a) provision of investment incentives, (b) provision of physical infrastructure and information networks in which small enterprises can operate efficiently, (c) measures to eliminate regulatory frameworks that constrain SSEs, (d) encouragement of institutions and organizations to provide soft loan lending to the SSE and *jua kali* sector,[16] and (e) provision of promotional programs such as training, counseling, and extension (Kenya, 1989, 1992b). The government, together with local urban authorities, has helped artisans to construct workshops (called "*nyayo* sheds").[17] More recently, the government has targeted vocational and middle-level training as a means of increasing entry and raising productivity in the SSE sector.

Specific measures are included in the government strategy to address the special constraints facing women in the small-scale enterprise sector (Kenya, 1992b). These include legal constraints, cultural barriers, limited access to credit, and low literacy rates (Keino, 1992). Labor force participation by women in the nonagricultural sector increased from 14.5% to only 20% between 1963 and 1984 in Kenya. Furthermore, women are disproportionally absorbed in the informal sector, where income generation is low and uncertain, and not in the mainstream of the modern sector.

The Local Authority Development Programme

Another recent policy initiative in Kenya concerns strengthening local authorities to enable them to provide competent administration and management and to increase their revenue-generation capacities. The LADP seeks to improve the capacity of local urban authorities to manage their urban development programs. Technical support and funding is provided by the German Agency for Technical Cooperation and the U.S. Agency for International Development. Recently, the program has initiated a number of urban development projects aimed at providing essential urban services and revenue generation for the urban authorities. The projects developed under this program include construction of marketplaces, bus parks, and slaughterhouses, as well as management training. It is expected that the LADP will strengthen local authorities' entrepreneurial capacity to initiate development activity.

To improve the financial standing of local government authorities, the government introduced local service charge levies to businesses and employees in 1988-1989.[18] In the past three years, the local service charge has become the dominant source of revenue for local authorities. The revenue from local service charges is used to improve existing services and to initiate new ones. Local authorities also derive revenue from other sources, such as rates, licenses, and property taxes. The significance of these measures is that they represent a major reversal of a trend evident since independence, whereby the central government had gradually usurped control over the disbursement of development funds from local government authorities while also undercutting their revenue base.

Institutional Changes and the District Planning Process

In Kenya, there is a continuing commitment to making the district planning process work effectively. The main challenges facing the process include (a) the need to strengthen subdistrict-level planning activities and empower local communities to become more involved in the development process; (b) human resource development requirements in the form of training district officials on methods of incorporating participatory, environmental, as well as technical and economic dimensions into the district planning process; and (c) strengthening of local government capacity to initiate development activity and provide entrepreneurial initiatives.

A number of institutional and policy measures designed to address these concerns are already being implemented. The government continues to offer training and professional development for district development officers. More recently, it has begun posting development officers at the subdistrict (divisional) level and also supports a large extension service. However, there is recognition in the government that the formal planning structure, its technical and tight scheduling requirements and the inevitably macro-level nature of the resulting districtwide plan, can inhibit local involvement. As a result, there is growing interest in the government's moving away from training in narrow traditional (procedural/technical) planning and toward forms of training in planning methods that incorporate participatory, environmental, equity, and sociocultural dimensions, as well as technical and economic dimensions (more holistic/participatory/action-oriented planning).

A major constraint lies in the latent resistance to change on the part of the establishment and the continuing tension between the conservative view, which favors technocratic, bureaucratic, and top-down planning, and those who see the need for the greater role of local government and local communities in development planning. The relation between central government and local authorities is still highly vertical with regard to decision-making processes. There is a need to reduce the extensive powers of the minister of local government (provided by the Local Government Act), which too often tend to stifle initiative within local authorities. This would give local authorities greater discretion to use more effectively their enhanced financial and revenue base and the scope to utilize the administrative and management skills fostered through the LADP.

The district planning process provides great potential for integrating the diverse programs currently pursued by the government, something that has been lacking before. For example, it is now well known that the economic well-being of local urban authorities is intricately linked with that of their rural hinterlands (Evans & Ngau, 1991). Yet district planning in Kenya has been very slow to respond to challenges facing rural areas, such as deterioration in the current farming systems, land degradation, and falling commodity prices and production levels in agriculture. Recently, many coffee and tea farmers in Kenya faced with declining prices have begun shifting to other crops and land uses and are increasingly organizing boycotts of coffee and tea picking.

Political Restructuring and Local Policy Activism

Since 1990, a strong pro-democracy movement has been raging in Kenya against a highly entrenched one-party regime.[19] The struggle led to the repeal of the constitution at the end of 1991, making Kenya a multiparty state. Within a year the country burst into multiparty activities, with no fewer than 12 registered and unregistered political parties.[20] More significant, political restructuring in Kenya has brought about the emergence of strong political and local policy activism. It has given rise to a wide range of local interests and pressure groups, including environmental groups, pro-democracy activists, party youth groups, legal and civil liberties organizations, labor/business organizations, and professional organizations, as well as a resurgence of ethnic, cultural, and religious movements.

The Kenya Law Society and various church organizations continue to play a significant role in advocating for democracy and civil liberties. Numerous legal defense and civil liberties organizations have emerged, such as the Committee for Democratic Change, the Public Law Institute, the Federation of Women Lawyers (Kenya Chapter), the Kenya Human Rights Commission, the Legal Advice Centre, and International Commission of Jurists (Kenya). These legal organizations are increasingly involved in matters pertaining to local policy, environmental conservation, and safeguards for urban consumer interests. For example, the Public Law Institute has successfully represented suits with the Kenya Consumer Organization against proposals by Kenya Posts and Telecommunication and Kenya Power and Lighting Company to raise the costs of utilities. It has also represented the Green Belt Movement in successfully barring the KANU Media Trust Corporation from building a multistory complex in the middle of Uhuru Park. In another instance, the Jevanjee family, working together with local environmental activists and the Kenya Architectural Association, successfully fought against an action to take over Jevanjee Gardens in the city center by a business corporation that intended to construct underground parking and a shopping complex. Increasingly, legal organizations are taking representative suits with environmental, professional, and neighborhood organizations against abuse of power by central government, local authorities, and industries dumping toxic wastes and polluting the environment.

Church organizations, including the National Christian Council of Kenya, the Church of the Province of Kenya, and the Roman Catholic church, have formed task forces such as the Justice and Peace Commission that have proven effective in championing the pro-democracy movement and defense of human rights through proclamations and pastoral letters. More recently, they have offered avenues for conflict resolution in interparty and intraparty feuding by organizing symposia. The church groups have also taken an active role against industrial pollution. One case is the ongoing protests led by the Catholic church in Thika against an industry in the town considered to be emitting toxic pollution endangering the health of a large nearby residential community. Increasingly, communities are becoming courageous in fighting to safeguard their environment and well-being, and community/religious-based bodies are emerging that are successfully providing for local conflict resolution.

Abuses of human rights through detention without trial, police torture, and disappearance of people perceived as political opponents were widely reported in Kenya from the early 1980s. Currently, pressure groups are waging campaigns for release of political detainees and an end to arbitrary arrests by police. Such pressure groups include the Release Political Prisoners Organization, Mothers in Action, Mothers Voice, Human Rights Commission, Kenya Civil Liberties Union, Kenya Human Rights Chapter, and Kenya Ex-Political Prisoners Organization. These organizations have successfully campaigned for the release of many political detainees, and have become a powerful force against human rights abuse.

A number of organizations have also emerged to advocate for the interests of women. Such organizations include the Legal Women Voters Organization, Mothers in Action, Mothers Voice, the Anti-Rape Commission, and the National Action Committee for the Status of Women. In the countryside there are numerous women's organizations, such as the Mothers Union, Maendeleo Ya Wanawake, the Green Belt Movement, and many women's *harambee*/social groups. These various groups are active in promoting the economic and social well-being of women through the initiation of local development projects and the building of community health clinics and water supply systems.

In the urban areas, there are organizations formed by local business entrepreneurs in the private sector to safeguard their business interests, too. Examples include the *jua kali* (informal sector), Kundi La Wasanii (an organization of artisans), the Nairobi Women Hawkers Association, and Nairobi Matatu (Minibus) Association. Their basic aims are to end harassment by the police and city commission officers and to seek better places to operate their businesses. Recently, when the Nairobi street vendors invaded the main city commercial streets, the commercial merchants also formed their own business organization, seeking to clear the street vendors from their business doorsteps. This led to a confrontation between the two groups, but the city commission intervened and worked out a settlement.

In the resurgence of current political and policy activism, two obstacles to change—fear and apathy—have been rapidly overcome among the people. Furthermore, there is increasing awareness among the people about their rights and roles at all levels of government. There are already organizations providing education to the society on matters of legal, political, environmental, and social awareness. These include the Legal Aid and Educational Programme, the Kenya Consumer Organization, A New Approach to Democracy, Alternative View, the Free and Fair Election Movement, the Middle Ground Group, the Moral Alliance for Peace, and the Legal Advice Centre.

What these numerous movements signify is the dawn of a new era of citizen empowerment that is destined to change further the relations among the central government, local government authorities, the market, and the civil society. It is certainly changing the way the development process (and ultimately development planning) is viewed and conducted. It has raised demands for greater participation of communities in the development and planning processes and concerns for sustainable development. What is needed is a reorientation of central government and local government institutions to allow local enthusiasm for participation in development and to channel it toward positive change.

Notes

1. This division is well illustrated by the shifting orientations found in the First Development Plan, 1964-1970 (Kenya, 1964), the Sessional Paper No. 10 of 1963/1965 (Kenya, 1965a), and the Revised First Development Plan, 1966-1970 (Kenya, 1966). The First Development Plan (referred to as the Red Plan) stated that Kenya would develop as a socialist/communist state, whereas the Sessional Paper No. 10 of 1963/1965 and the Revised First Development Plan (called the Green Plan) changed Kenya's economic orientation to market capitalism. Both documents emphasized private ownership of means of production and foreign investments.

2. At independence, 1963, there existed a Directorate of Planning in the Ministry of Finance and Economic Planning. In December 1964, a separate Ministry of Finance (MF) and Ministry of Economic Planning and Development (MEPD) were created. In 1970, the two ministries were merged into one, the Ministry of Finance and Planning (MFP). In 1978 they were split again into the Ministry of Finance and the Ministry of Economic Planning and Community Affairs (MEPCA), later renamed the Ministry of Economic Planning and Development in 1980. They were merged again and called MFP in 1983, and since 1985, the MFP and MEPD have remained separate ministries.

3. In Kenya, districts and the county councils (local authorities) actually share the same geographic boundaries, one representing the government's central administrative presence at the district (headed by the DC) and the other representing the local government with elected officials (headed by the county council chairman, or mayor, in the case of urban authorities).

4. The transfer was done in October 1969, through the Local Government Transfer of Functions Bill (see Gertzel et al., 1969). In transferring the responsibilities, the central government accused local authorities of financial weakness, incompetence, and petty power rivalry.

5. *Harambee* is a slogan and a rallying self-help motto used in Kenya to mean "Let us pull together." The term was popularized by the country's first president, Mzee Jomo Kenyatta.

6. Development Advisory Committees (DAC) were created by the Revised First National Development Plan, 1966-1970. The First National Development Plan, 1964-1970, had created similar committees called Community Development Committees

(CDCs) for local community planning. The CDCs became subcommittees of DACs, but the latter never became operational because DDCs were not functioning. The CDCs exist today and concern themselves mainly with local social services.

7. A rather long period (1966-1970) was spent surveying and identifying the first SRDP pilot projects, developing plans, and applying for funding. The Kenyan government was reluctant to commit funds to what appeared to be risky projects with uncertain payoffs. Although it was a brief experiment, several innovations were provided by the SRDP. The establishment of an overall coordinating committee for SRDP was the predecessor of the present-day DDCs in Kenya. The donor aid program that enabled it led to the establishment of the Rural Development Fund and the EEC funds for micro-level projects.

8. The RDF was formed by bringing together two funding sources—one called District Development Grants, initiated in 1971 to fund a variety of small locally identified projects with large self-help components, and another called the Rural Works Program, designed to fund small, labor-intensive projects.

9. Kenya's decentralization is therefore what is referred to as *deconcentration,* rather than devolution. It essentially empowers administrative officers at the lower government echelons, instead of the local people.

10. The extensive powers of chiefs in Kenya are given by a constitutional act called the Chiefs Act, which dates from the colonial era, when it was established to deal with natives during the period of emergency.

11. The table is based on data contained in annual annexes (1985-1986) to District Development Plans. All DDCs are required, in the District Focus Strategy, to indicate origins/sources, type of project, and priority ranking of all the projects they submit. Priority ranking is provided within three categories: high priority, medium priority, and low priority.

12. Districts tend to produce long shopping lists, ostensibly to draw more national resources.

13. Structural adjustment measures introduced in early 1980 included (a) reduction of government expenditure on social services, (b) increased taxation, (c) charging users for services they receive (so-called cost sharing), (d) currency devaluation, (e) privatization of state-owned enterprises, and (f) increases of producer prizes in agriculture. Since 1990, the new generation of SAPs include a condition of good governance.

14. The RTPC program can be distinguished from the previous growth center strategy in that it (a) includes specific provision for funding and targets a reasonable and attainable number of centers over a long period of time for implementation, (b) has a clear set of selection criteria and guidelines to facilitate prioritization, and (c) has targeted increased economic activity in both site selection and project package provision.

15. The term *small-scale enterprise* in Kenya is used in a broad sense to refer to enterprises consisting of up to 50 employees. The small-scale enterprise sector comprises a range of enterprises, including self-employed artisans, *jua kali* enterprises having a few employees (see note 16), cottage industries, sole proprietors, and small enterprises in the formal business sector having some 10 or more employees. These small enterprises may be engaged in trade, services, and/or manufacturing.

16. *Jua kali* refers to the businesses of artisans, traders, and services conducted in open-air places in many urban places in Kenya. The term literally means "hot sun."

17. *Nyayo* is a slogan and expression used in Kenya to underline the principles of peace, love, and unity. The term literally means "footsteps," and has been popularized by President Daniel Moi, who succeeded President Kenyatta.

18. The initial rate at which local service charges are levied was set by the central government. However, there are cases now where some local authorities have introduced additional charges.

19. Kenya's one-party political system goes back to 1964. Between 1964-1966 and 1969-1982, Kenya was a de facto one-party system during the time of President Jomo Kenyatta (1963-1978) and President Daniel Moi, who succeeded Kenyatta in 1978. In the brief period 1966-1969, an opposition party existed called Kenya Peoples Union, led by Oginga Odinga, the current chairman and presidential candidate of Forum for Restoration of Democracy (FORD-Kenya). In 1982, the KANU government under President Moi led Parliament to amend the constitution, making Kenya a de jure one-party state. After a protracted struggle by the pro-democracy movement, the government changed the constitution in December 1991 to allow for multiparty democracy.

20. These include KANU, led by Mr. Daniel Moi; FORD-Kenya, led by Mr. Oginga Odinga; FORD-Asili, led by Mr. K. Matiba and Mr. M. Shikuku; the Democratic party, led by Mr. Mwai Kibaki; the Social Democratic party, led by Johnstone Makau; the Kenya Social Congress, led by Mr. Anyona; the People's Party of Kenya, led by Mr. Samuel Mwaura; the Kenya Democratic party, led by Mr. Ngonya wa Gakonya; the Green African party, led by Mr. Godfrey Mwireria; the Islamic Party of Kenya, led by Mr. Mohammed Abdullahi; the Kenya National Congress, led by Mr. Titus Mbathi; and the Party for Independent Candidates of Kenya, led by Mr. John Mwau. Youth organizations have also emerged and serve as fronts for the above parties. So far the following have been established: Youth for KANU92, associated with KANU; Operation Moi Win, associated with KANU; FORD-Youth, associated with FORD; Association of Young Professionals, associated with the Democratic party; and Kenya Youth Foundation, associated with the Kenya Democratic party. There is considerable inter- and intraparty feuding among the new political parties. They, like KANU, have done little to explain to the electorate their economic and social policy programs. Because of their fragmentation and internal feuding, there was no clear winner in the national elections in Kenya in late 1992; President Moi continued in office.

References

Chege, M., & Kimura, J. (1982). *Budget estimate preparation at district and provincial levels: A case study of the ministries of agriculture and livestock development in Meru District, Eastern Province.* Report prepared for the Management Systems Unit, Ministry of Agriculture, Nairobi.

Cohen, J., & Hook, R. (1986). *District development planning in Kenya.* Unpublished manuscript, Ministry of Planning and National Development, Rural Planning Division, Nairobi.

Delp, P. (1980). *District planning in Kenya* (Development Discussion Paper No. 95). Cambridge, MA: Harvard Institute for International Development.

Evans, H. (1988). *National urban policy in Kenya: Past experience and new directions* (Development Discussion Paper No. 272). Cambridge, MA: Harvard Institute for International Development.

Evans, H., & Ngau, P. M. (1991). Rural-urban relations, household income diversification and agricultural productivity. *Development and Change, 22,* 519-545.

Gaile, G. L. (1992). Improving rural-urban linkages through small town market-based development. *Third World Planning Review, 14,* 131-148.

Gertzel, G., Goldschmidt, M., & Rothchild, D. (Ed.). (1969). *Government and politics in Kenya.* Nairobi: East African Publishing House.

Heyer, J., Ireri, D., & Moris, J. (1971). *Rural development in Kenya.* Nairobi: East African Publishing House.

Hunt, D. (1984). *The impending crisis in Kenya: The case for land reform.* Brookfield: Gower.

International Labor Organization. (1972). *Employment, incomes and equity: A strategy for increasing productive employment in Kenya.* Geneva: Author.

Keino, I. C. (1992). *Participation of women in small scale enterprises in Nairobi: A study of Wakulima, Burma and Kariokor market areas.* Unpublished master's thesis, University of Nairobi, Department of Urban and Regional Planning.

Kenya, Republic of. (1964). *Development plan, 1964-70.* Nairobi: Government Printer.

Kenya, Republic of. (1965a). *African socialism and its application to planning in Kenya* (Sessional Paper No. 10 of 1963/5). Nairobi: Government Printer.

Kenya, Republic of. (1965b). *Local government commission of inquiry, 1965: Hardacre Commission report.* Nairobi: Government Printer.

Kenya, Republic of. (1966). *Development plan, 1966-70.* Nairobi: Government Printer.

Kenya, Republic of. (1971). *Public Service Structure and Remuneration Commission: Report of the commission of inquiry* (Chairman, D. N. Ndegwa). Nairobi: Government Printer.

Kenya, Republic of. (1984). *District focus for rural development* (rev. version). Nairobi: Government Printer.

Kenya, Republic of. (1986). *Economic management for renewed growth* (Sessional Paper No. 1 of 1986). Nairobi: Government Printer.

Kenya, Republic of. (1989). *A strategy for small enterprise development in Kenya: Towards the Year 2000.* Nairobi: Government Printer.

Kenya, Republic of. (1992a). *Economic survey, 1992.* Nairobi: Government Printer.

Kenya, Republic of. (1992b). *Small enterprise and jua kali development in Kenya* (Sessional Paper No. 2 of 1992). Nairobi: Government Printer.

Livingstone, I. (1981). *Rural development, employment and incomes in Kenya.* Addis Ababa: International Labor Organization, Jobs and Skills Programme for Africa.

Makokha, J. (1985). *District focus: Conceptual and management problems.* Nairobi: Japrints.

Ngau, P. M. (1987). Tensions in empowerment: The experience of the Harambee (self-help) movement in Kenya. *Economic Development and Cultural Change, 35,* 523-538.

Ngau, P. M. (1989). *Rural urban relations and agrarian development in Kutus Area, Kenya.* Unpublished doctoral dissertation, University of California, Los Angeles, Graduate School of Architecture and Urban Planning.

Ngau, P. M., & Gaile, G. (1992). *Measures for strengthening the implementation of Rural Trade and Production Centres programme (RTPC) in Kenya.* Nairobi: University of Nairobi, Housing Research and Development Unit.

Ondiege, P. O., & Aleke-Dondo, C. (1991). *Informal sector assistance policies in Kenya.* Nairobi: University of Nairobi, Department of Urban and Regional Planning.

Oyugi, W. O. (1981). *Rural development administration: A Kenyan experience.* Nairobi: Vikas.

World Bank. (1990). *World development report: Poverty.* New York: Oxford University Press.

10

The New Localism From a Cross-National Perspective

EDWARD G. GOETZ

The chapters presented in this collection describe a set of fundamental relationships and processes that are occurring in urban settings across a wide range of locations and in the context of quite disparate political economies. They provide evidence of rather fundamental political restructuring occurring as a result of the economic dislocations experienced since the mid-1970s. This political restructuring has two dimensions: a vertical dimension that includes the ways in which central and local governments are rearranging their relationships with each other, and a horizontal dimension that includes the redefinition of responsibilities between the public and the private sectors at the local level. That is, for the cases described in this book, local governments are experiencing fundamental changes in the division of authority and responsibility vis-à-vis the center or vis-à-vis the market, or in some cases both.

Further, the contributors to this volume discuss the implications of this restructuring for the issues of participation and equity in development. An outline of the relationships is shown in Figure 10.1. Though the existence of both vertical and horizontal restructuring is a constant throughout the chapters, the interrelationships among the variables highlighted in Figure 10.1 are complex and quite dependent on localized political factors. It is the very contextual nature of these differences that demonstrates the importance of local political factors.

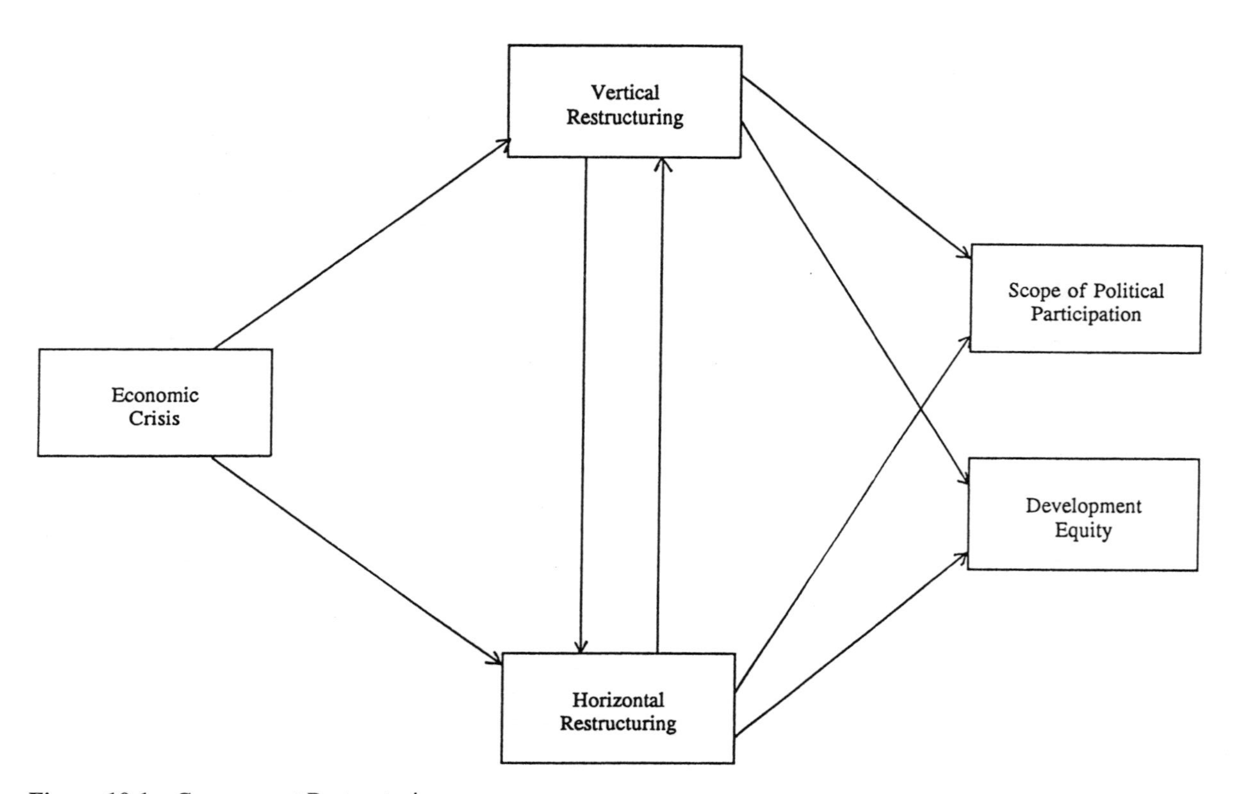

Figure 10.1. Government Restructuring

The following discussion will examine the case studies presented in the earlier chapters for the commonalities they reveal about (a) the impact of economic crisis and restructuring on the rearrangement of local politics both vertically and horizontally, (b) the interrelationship between center-local relations and public-private arrangements, and (c) the impact of these political changes on the scope of participation at the local level and equity in development.

Where the argument so indicates and the data permit, the discussion of the new localism to follow will be illustrated by reference to a cross-national data set on local fiscal issues conducted by the Fiscal Austerity and Urban Innovation (FAUI) Project (see Clarke, 1989). The FAUI Project is centered at the University of Chicago and incorporates the data collection efforts of scholars around the world. Surveys of mayors, chief administrative officers (CAOs), and city council members have been completed for an exhaustive list of cities in 12 countries. The survey data have been matched by fiscal and demographic data for the cities in the sample.[1] The use of this unique data set will allow a more extensive examination of the relationships described in the chapters of this book. The scope of the data collection effort undertaken by FAUI will allow the discussion of the new localism to overcome problems of generalizability that are inherent in the case study approach. Case studies are able to achieve a high degree of "internal validity" in that analysis is typically quite contextual and examination of interactions among many factors is possible. The case study generally provides a high degree of confidence that the nature of relationships and processes are well understood in the research setting. At the same time, reliance on a single case does not allow the analyst to assess how robust the findings are—that is, how well they apply to other settings, at other times, with different actors. Cross-sectional survey data, on the other hand, allow for testing relationships across widely divergent settings and thus provide a very useful complement to the case studies.

Political Restructuring Defined

All of the chapter authors trace political restructuring back to economic crisis, dislocation, or extreme fiscal strain. As Thomas Klak and Jamie Rulli argue in Chapter 7, on the Caribbean nations, local policy responds to global economic pressures as well as to pressures from large and powerful economic interests. In addition, they point out that national policy is also quite responsive to global economic cues. Shifts in regimes of accumulation have important impacts on how national and local governments organize and direct economic policy, including the division of labor between central and local governments, and the degree of economic activism and intervention pursued by local governments. In the following subsections I will examine the attributes of political change at the local level.

Vertical Restructuring

As Dele Olowu argues for Nigeria (Chapter 8), *vertical restructuring* is an attempt to rationalize center-local relations that have been made irrational by either economic change or new political circumstances. Center-local relations are, of course, critical to both local and central officials. The rearrangement of these relations is therefore potentially the result of top-down efforts of central officials or bottom-up pressure of local government authorities. Rather than static, institutional relationships, center-local relations are fluid, political settlements that are negotiated and frequently renegotiated. Bottom-up pressure can come about because of the concern for equity in public and social services across regions, or because local officials orient their politics toward increasing central support during periods of fiscal strain (Markusen, 1989). Similarly, central government officials can devolve program responsibility to local levels, rearrange fiscal burdens, and thereby resolve their own budgetary problems.

The dimensions along which vertical restructuring occurs relate to the scope of autonomy for local governments in pursuing economic initiatives, the degree of initiative allowed in fiscal matters, and even the latitude granted in the implementation of central government programs. In Chapter 5, Wisla Surazska argues that for Poland the relationship between center and local governments is a debate of corporatist and communitarian political ideals. After the fall of the government, the adoption of a mode of center-local relations was tilted in favor of a more decentralized design in part because local elections preceded the formation of national democratic institutions. The first expression of a more participatory civil society in Poland was through the activity of local governments.

Yet in Poland and Hungary the legacy of decades of strong centralized rule has not vanished quickly. Gábor Péteri reports in Chapter 6 that despite being somewhat less interventionist, the central state in Hungary has not diminished much if measured by government expenditures. Likewise in Poland, the new democratic processes that emerged at the local level have receded as participation in local elections has dropped to low levels.

The Eastern European cases together with the African cases suggest that vertical restructuring is only partially achieved by the actions of the central government in formally or informally reforming the institutions of government. What is also necessary for a more lasting and meaningful realignment of internal political and administrative activities is the formation of effective institutions and political mechanisms at the local level. As Olowu points out for Nigeria, the central military authorities carried out a series of important and far-reaching reforms in the federalist structure of that country. Yet, the emergent local governments, though noticeably more important and active than their predecessors, still failed to play a significant role in the civic life of local areas. Instead, alternative organizations, community development associations, sprang up to fill the vacuum and provide the means for more widespread political participation at the local level. In Kenya, reforms out of the center produced widespread political activity and the emergence of new political parties and interest groups, as well as community-based organizations concerned with civil liberties, ethnic and cultural matters, and the environment (see Ngau, Chapter 9).

Horizontal Restructuring

Horizontal restructuring refers to the redefinition of public/private and the reconstitution of local government roles vis-à-vis the market. In fact, this type of restructuring can take many forms. The chapters on Great Britain, for example, suggest that there are both "left" and "right" ways of achieving horizontal restructuring.

The experiment of municipal socialism undertaken by a number of local governments in Great Britain during the early 1980s involved extensive reworking of the lines between public and private sectors. According to James Chandler (Chapter 3), local governments working under this paradigm sought to increase public control over investment decisions, to create new forms of social ownership of economic assets, to develop strategies for industrial development, and to provide support for unions, women workers, and socially useful products. Municipal socialism in Great Britain involved a dramatic extension of public authority over private investment. Cities used whatever means were available to them to promote, and even engage in, economic activity of a "socially useful" sort.

To adapt somewhat the framework provided by Péteri in Chapter 6, there are four roles that local governments may take with respect to the market. The first is a fully *participatory* role, in which the local government acts as a market entity. This includes the ownership of productive assets as noted by Péteri, but also includes the participation of government in equity positions and partnership arrangements. This is the "entrepreneurial" type of activity that Hungarian cities have already adopted for themselves, quite akin to the actions of Western cities during the past 10 years. It is of note that the Eastern European cities have done this in the face of the economic advice of free-marketeers from the West who have gone east to oversee the economic transformation. This first role also implies an element of planning for local economic development and the purposive use of government resources to direct the local economy.

The second role might be called a *facilitative* role, à la Jacobs (Chapter 4), in which the local government uses more conventional methods to facilitate economic activity. This can include the use of subsidies and incentives to encourage economic development, but also includes the delivery of public services that serve the same purpose, such as infrastructure.

The third role is a *regulatory* one, in which the local government uses the tax system and regulatory systems to monitor the private market. This is a traditional role for local governments in many countries.

Finally, another role is one that might be called *adjustive,* in that the local government is involved in the provision of social and public services that are in place to mitigate the effects of the market, or to ameliorate the uneven economic distribution of market systems. As Surazska points out, local governments in Poland have been given the responsibility of dealing with the social "aftermath of the collapse of the centrally planned economy." This social service role is also a more traditional role for local governments.

The horizontal adjustments being made by local governments involve movement among these different roles, especially between the first two and the others. There is a great deal of variation in how these roles might be played. Richard Hula (Chapter 2) and Péteri point out that there is a high potential for privatization of public services (provided under the second and fourth roles described above) that also affects the degree of restructuring being assessed. Chandler argues that city officials move from the fourth role to the first in an attempt to have a more meaningful say in determining local economic outcomes. In fact, in the United States, the great reorientation of urban policy between the 1960s and the 1980s was a shift from antipoverty social service programs (the fourth role) to a more interventionist posture through economic development programs aimed at influencing patterns of economic activity and capital investment strategies of the private sector. In Great Britain, the Thatcher government introduced reforms that were aimed at moving cities down the scale from the highly interventionist techniques of municipal socialism to a more facilitative role in economic policy dominated by private sector actors. Brian Jacobs indicates that after the Thatcher reforms and the end of the municipal socialism experiment, the city of Birmingham spent greater effort in social amelioration than it did in economic revitalization.

Cross-National Data on Restructuring Attitudes

The types of political restructuring strategies described above are widespread. The cross-national FAUI data allow an investigation of the orientation of local officials when considering fiscal problems and their solutions. CAOs were asked to specify the sources of fiscal problems and the types of solutions their cities have used. The responses indicate the degree to which problems and solutions are identified with respect to center-local or public-private relations, compared with internal factors. Table 10.1 displays the mean responses of CAOs surveyed. Responses were coded from 0 to 100, indicating the importance of each problem and the degree to which specific solutions have been tried. The data show that city officials are quite likely to blame the center for fiscal problems. The mean responses for individual items show that cities consider the loss of revenue from central- and state/provincial-level governments most important, matched only by problems related to inflation. Also important to CAOs is the cost of centrally mandated programs. When the individual items are aggregated to reasons related to the economy, local political issues, and center-local relations, center-local relations are indicated to be significantly more important problems

Table 10.1 The Source of Fiscal Problems

Source of Fiscal Problem	$\bar{X}$	n
Loss of central government revenue	50.0	1,289
Loss of revenue from state/province	57.3	1,022
Centrally imposed tax/revenue limits	31.6	1,112
Centrally mandated costs	46.9	1,310
Summary variable: Center	47.7	748
Inflation	56.0	1,453
Unemployment	38.6	1,421
Declining tax base	43.9	1,411
Increasing service demand	47.0	1,451
Summary variable: Economic	46.1	1,345
Local pressures to reduce taxes	32.8	1,295
Failure of bond referenda	13.5	1,045
Pressure from municipal employees	34.6	1,439
Summary variable: Local/Political	24.8	1,026

NOTE: The survey question read, "In the last three years, how important have each of the following problems been for your city's finance?" Answers are scaled from 0 (least important) to 100 (most important).

than either of the other two. This is rather significant, given the presumption that economic conditions might present themselves as the most obvious cause of local fiscal problems.

Similarly, CAOs were asked to indicate how they were dealing with fiscal problems. Officials were given a list of fiscal strategies and asked which were most important to them. The analysis below categorizes responses by the impact of the strategy on the size of the public sector. In the first category are strategies that shrink the government, such as the elimination of programs and reductions in the public workforce. In the second category are fiscal strategies that increase the size and scope of the public sector, such as tax increases and pursuing additional intergovernmental resources. The third category consists of those generally administrative procedures that are neutral with respect to the size and scope of government activity, such as productivity improvement through management techniques, lowering surpluses, and deferring payments.[2]

Significantly, the mean response to the second category (answers that indicated a *growth* in the size and scope of government) was highest among the three categories, 29.7. The mean response for the neutral category was 28.6, and the mean response for answers indicating a reduction in the public sector was 25.9. These data clearly indicate that even in the face of fiscal problems, and even when considering actions to ease fiscal concerns, local officials were equally compelled, indeed slightly more compelled, to follow strategies that expanded the scope of government rather than shrinking it. This tendency is greatest in France and least pronounced in the United States.[3] The findings suggest a strong commitment across cities in the sample to local government activism in economic and fiscal issues, especially in the face of the often-stated assumption that the most common response to fiscal crisis is to reduce government size.

Of the many types of fiscal strategies rated by respondents, a few can be identified as attempts to reorganize directly either center-local or public-private relations. Table 10.2 lists those strategies with direct implications for horizontal or vertical restructuring. The mean responses associated with pursuing intergovernmental assistance, selling assets, and contracting out put these strategies in the mid-range compared with all other strategies reported.

One-third of the cities in the FAUI sample report that fiscal strategies involving the reorganization of city/market boundaries are at least "somewhat important." The issue of public ownership of assets, which is at the heart of the debate in Poland and Hungary, is "somewhat" or "very important" in 33% of the sample cities. The contracting out of public responsibilities to private sector actors is "somewhat" or "very important" in 36% of the cities. The use of these strategies is relatively greater in the Japanese, Finnish, and Australian cities in the sample. Cities in Norway and Canada seem to put less emphasis on these strategies.[4] Some 45% of the sample cities report that the pursuit of additional intergovernmental resources is at least somewhat important. The use of other vertical strategies is more limited, however, taking on importance in generally less than 25% of the cities. In all, the FAUI data show a significant commonality in the issues faced by cities across the sample and in the strategies they employ.

The data presented in Tables 10.1 and 10.2 in all likelihood underrepresent the degree to which cities are undertaking new market roles and new roles in relation to the center, for the data do not explicitly deal with development policy, where such roles are most evident. Given

Table 10.2 The Use of Vertical and Horizontal Fiscal Strategies

Fiscal Strategies	$\overline{X}$	n
Strategies with implications for center-local relations		
Get additional intergovernmental resources	32.8	1,455
Shift responsibilities to other governments	11.8	1,133
Contract out to other governments	16.6	1,028
Decrease services funded by intergovernmental revenues	18.2	1,294
Strategies with implications for state-market relations		
Sell assets	26.1	1,453
Contract out to private sector	25.7	1,382
Joint purchase agreements	18.7	1,298

NOTE: The survey item read, "Please indicate the importance in dollars of each fiscal management strategy listed below." Answers are scaled from 0 (least important) to 100 (most important).

the greater costs, less certainty, and greater permanence of vertical or horizontal fiscal strategies, the data indicate a surprising willingness on the part of local governments to restructure the economic policy arena. The survey data support the contention made by the case studies presented here about the importance of changing political roles at the local level.

Economic Crisis as the Origin of Political Reorganization

Impact of Economic Crisis on Vertical Restructuring

In many Western democracies the central government has typically intervened during periods of economic decline in order to assist hard-hit regions (Judd & Parkinson, 1990a, 1990b). Generally, this was achieved through centrally funded and centrally controlled programs of urban renewal. The preceding chapters, however, in the main tell a story of central government reforms that put greater latitude and economic responsibility in the hands of local officials.

In the African states of Kenya and Nigeria, rapidly declining growth rates led to center-local reforms during the 1970s. In both cases the reforms enhanced the position of local governments by allowing them greater powers. In both cases, however, the reforms did not take hold, in the sense that little local initiative took place. During the 1980s, as the structural adjustment programs began to have adverse effects on the local economies, and as Nigeria suffered from the collapse of oil, more widespread modifications in center-local relations were made, and more effective forms of local participation were created. It is important to note, of course, that regime change is also quite important in spurring a reassessment of center-local relations. In the Nigerian case, the military regime also pursued decentralization for overtly political reasons. Greater democratization and a transition to civilian rule required greater devolution of political activity to the local level.

In the Eastern European cases, it might be argued that the fall of Communist party governments was precipitated by economic crisis, and therefore the resultant changes in intergovernmental relationships are to be explained, ultimately, by economic factors. In fact, one need not resort to that indirect argument. In Poland the link between extreme deficits faced by the central government and the structural readjustment of center-local relations was direct. As Surazska points out, the central Polish government unloaded a portion of its debt to the localities as part of its reform of the intergovernmental system.

The FAUI data allow some investigation of the link between economic crisis and vertical restructuring. In the U.S. context, the Markusen (1989) hypothesis suggests that local governments in severely depressed and declining regions orient their politics toward the federal government, hoping to elicit a rescuing response from the center. The FAUI data can be used to test that hypothesis in cross-national perspective. Measures of fiscal and economic stress were correlated with the use of vertical restructuring strategies described earlier. The results, listed in Table 10.3, show some support for that hypothesis. Unemployment levels show moderate but consistent correlations with the vertical strategies examined (except for contracting out to other governments, for which the direction of the relationship is negative). Declining population levels were correlated with the shifting of program responsibilities to other levels of government. There was no relationship between per capita debt and any of the vertical fiscal strategies.

These relationships were not changed significantly by taking the U.S. cities out of the analysis. Thus, though Markusen explicitly makes her argument for the U.S. setting, the cross-national survey data indicate the relationship is robust even across diverse national settings.

Table 10.3 Correlations Between Economic Distress and Vertical Fiscal Strategies

	Population Change[a]	*Change per Capita Debt*[b]	*Change in Unemployment Rate*[c]
Get additional intergovernmental resources	.00	–.03	.05*
n	1,207	741	1,101
Shift responsibilities to other governments	–.10***	–.01	.20***
n	876	491	769
Contract out to other governments	–.03	–.01	–.12***
n	771	385	775
Decrease services funded by intergovernmental revenues	–.01	.02	.18***
n	1,050	664	943
Summary variable: Vertical	–.05	–.03	.10***
n	756	373	760

NOTE: Figures are Pearson product-moment correlations.
a. Figures on population change use slightly different time periods for the five countries for which data are available. For the United States and Japan the time span is 1980-1985; for Finland, 1981-1985; for Norway, 1981-1984; and for France, 1977-1982.
b. Data on change in per capita debt: United States, 1980-1985; Finland and Japan, 1981-1985; France, 1982-1985.
c. Data on unemployment rate: United States, 1980; France, 1975; Finland and Norway, 1981.
$^{*}p < .05$; $^{***}p < .001$.

Impact of Economic Crisis on Horizontal Restructuring

The Western cases show rather direct effects of economic crisis on the reconceptualization of public-private relationships at the local level. Jacobs argues that economic restructuring and political restructuring were intimately linked in Birmingham. With the rest of the region suffering from massive plant closures and economic redundancy, local officials felt the need to establish the local capacity to direct economic activity. Chandler documents similar processes for the rest of Great Britain. The very same process has driven new forms of economic development policy in Canada (see, e.g., Leveillee & Whelan, 1990), Germany (Mayer, 1987), France (Body-Gendrot, 1987; Le Gales, 1990), and the United States (Beauregard, 1989; Logan & Swanstrom, 1990; Smith & Feagin, 1987).

The Relationship Between Vertical and Horizontal Restructuring

Some of the more intriguing issues addressed in the previous pages center on the mutual interactive effects between the two types of political restructuring. In many of the cases presented in this volume, there were direct impacts of center-local realignment on the public-private relationships of local governments. This is perhaps to be expected, as local governments adjust to changes in mandated responsibilities by experimenting with new forms of policy making. Less expected, however, are the ways in which the actions of local governments, in redefining their orientation toward market processes, produced a response by the central government and, in the case of Britain, a subsequent adjustment of center-local relationships.

The impacts of center-local reforms on horizontal restructuring are fairly direct; in some cases they are designed. Hula notes that in the United States, federal policy shifted not only in regard to the amount of intergovernment support provided, but also in how those programs are to be implemented. Mandates for the privatization of certain programs obliged local governments to incorporate the participation of the private sector through contracting out and through public-private partnerships. The impact of federal program oversight and administration on the form of local policy in the United States has been noted by others as well, especially in the area of economic development (e.g., Clarke & Rich, 1985). In a similar manner, the Thatcher government attempted to alter the manner in which local governments in Britain were relating to the market by reining in the more activist governments. Interestingly enough, this action was predated by the creation of metropolitan councils by the Heath Labour government. These regional authorities used direct forms of economic intervention (the same forms objected to by the Thatcherites) at least partially as a way of creating a niche for themselves and establishing their own importance. Thus even portions of the municipal socialism experiment in Great Britain can be traced to shifts in center-local relations. The Western cases provide examples of how the accommodations between public and private sectors at the local level are molded by center-local relations.

Less anticipated, perhaps, are the ways in which changes at the local level affect political assessments made at the center. Municipal socialism as practiced in Great Britain during the early 1980s was contrary to the Thatcher government's economic ideas, which stressed an unfettered private sector engineering growth through the freer action of producers and investors. As well described by Chandler, the center was able to impose its own version of economic relations on local governments in Great Britain. The Thatcher government created new institutions to usurp the role of more radical local councils, enacted programs that directly decreased the market restrictions imposed by localities, and in notable cases directly abolished metropolitan governments and other mechanisms used by localities to pursue their "socialist" development schemes. Thus, in Great Britain, the reorientation of local public-private relations introduced by the municipal socialism movement induced a dramatic and widespread vertical restructuring by the central government. Interestingly, the same center-local strain is appearing in Poland, where local governments are not proceeding with privatization at the rate that the center prefers. Local governments are finding that the ownership of assets, and the revenue generated therefrom, is quite valuable in a new era of local fiscal responsibility.

Elsewhere, the effect of local restructuring on the actions of the center are, if somewhat less emphatic, still significant. Both the Kenyan and Polish cases show that the level of activism and political participation unleashed can inhibit the center from making further adjustments in center-periphery relations. In Poland, the emergence of local activism prior to the establishment of national institutions forced the hand of the central government in the reformation of the territorial system. In Kenya, Peter Ngau suggests, center-local relations could enter a new era because of the level of local political activism.

The FAUI data also provide support for the contention that vertical and horizontal political reorganization are often linked. The fiscal strategies examined earlier are generally not pursued in isolation from each other. That is, the use of both horizontal and vertical fiscal strategies is moderately and positively correlated for the cities in the FAUI sample (see Table 10.4). For example, the correlation between the two summary variables indicating reliance on vertical and horizontal strategies is .37 ($p < .001$).

Table 10.4 Correlations Between Vertical and Horizontal Fiscal Strategies

	Sell Assets	*Contract Private Sector*	*Joint Purchase Agreements*	*Summary Variable: Horizontal*
Get additional intergovernmental resources	.26***	.16***	.30***	.27***
n	1,442	1,366	1,289	1,358
Shift responsibilities to other governments	.13***	.19***	.16***	.22***
n	1,131	1,130	1,130	1,128
Contract out to other governments	.14***	.10***	.03	.16***
n	1,020	1,021	1,021	1,016
Decrease services funded by intergovernmental revenues	.37***	.35***	.39***	.46***
n	1,286	1,208	1,289	1,203
Summary variable: Vertical	.25***	.28***	.26***	.37***
n	1,012	1,011	1,012	1,010

NOTE: Figures are Pearson product-moment correlations.
***$p < .001$.

The Politics of the New Localism

Local Political Activism

One of the important local impacts of political restructuring is a broadening of participation and policy activism. As Susan Clarke indicates in the introduction, a major characteristics of the new localism is economic policy activism. At the center of both the vertical and horizontal restructuring described in this book is a greater role for local governments and nongovernment groups in economic and development matters. As the development policy arena opens up and more authority is ceded to local governments on issues of production, political opportunities also increase.

The changing nature of center-local relationships creates the space for new groups and new collections of interests to become involved in economic policy (see also Parkinson, 1991). The devolution of development authority creates a political vacuum that is filled by local political entrepreneurs (see, e.g., Schneider & Teske, 1992) or by new forms of political interests. This is especially true in the African and Eastern European cases, where the lack of prior political activity is more noticeable. In some cases, as in Kenya and Nigeria, increased political participation is one stated objective of center-local reforms.

In Poland, political participation increased at least for the short term in the form of the Citizen's Committees. These bodies were a check on both the public and the private sectors during the transitional period directly following the fall of the Communist regime and thus played an important role in the creation and expansion of civil society. Another group of actors emerging from political restructuring in Poland were the local authorities, voluntarily and self-consciously organized. As Surazska points out, these organizations quickly exceeded their mandate by moving from economic cooperation into political activities aimed at influencing decisions at the center. The Nigerian and Kenyan cases provide further support for the hypothesis that the devolution of program responsibility creates an environment conducive to group formation and greater levels of political participation.

The generally positive effect of devolution on participation is sometimes mediated by the type of local government response. For example, the privatizing pattern of development pursued in Baltimore and in Birmingham (at least after 1985) may have introduced new groups into the process, but these groups were not widely representative or particularly participatory. Hula, Jacobs, and Klak and Rulli all argue that economic policy making in their case studies narrowed the field of interests by allocating too much authority and responsibility to narrowly conceived private interests. In fact, the newly formed "quasi-public" bodies in Baltimore were able to avoid public scrutiny because of their equal status as "quasi-private" bodies. Hula argues that this could significantly narrow the range of policy options considered by restricting the input of nonbusiness interests. In Birmingham, Jacobs documents a shift of power away from the council and toward the corporate sector, though he points out that the business sector was not uniform in its expressed development interests.

Yet shifting public and private roles need not inevitably lead to a narrowing of participation. In many cities the redefinition of public and private responsibilities creates political conflict that brings new actors into the arena. In the extreme this is exemplified by the Birmingham riots, but it is also seen in the proliferation of nonprofit organizations and community-based groups in the United States (Giloth & Mier, 1989).

The FAUI data allow an examination of the degree of participation by nongovernment groups. Table 10.5 summarizes the participation of different government and nongovernment groups in fiscal policy making. The figures in column 1 represent the overall mean response to a question about how active each group is in the policy-making process. Columns 2 through 10 break down responses by country. As shown in Table 10.5, the most active groups are government actors. This is, of course, to be expected. The second group of actors, however, receiving only slightly lower scores, are all nongovernment groups. Public employee unions, business groups, neighborhood groups, and, interestingly, the elderly all play significant roles in fiscal policy making. The FAUI data show that a wide range of groups are playing important roles in local fiscal policy.

There is some evidence for the proposition that participation is related to wider experimentation with public and private roles. The countries in which cities reported a higher degree of horizontal fiscal strategies (Finland and Japan) are shown in columns 2 and 3; the countries in which cities reported lesser use of those strategies follow, in columns 4 through 7. Cities in the final three countries listed in Table 10.5 did not provide data on fiscal strategies. There is a slight tendency for the means for nongovernment groups to be higher in those countries where horizontal strategies are in greater use. The exception to this is Australia, where cities reported a fairly high use of horizontal strategies but indicate relatively low participation of nongovernment groups. Nevertheless, there is a suggestion in these data that redefining city-market relations is correlated with a higher degree of group political activity.

Development Equity

Distributional equity has not generally been considered a primary issue in the analysis, or the implementation, of local development policy. Nevertheless, several of the cases presented here suggest interesting relationships between political restructuring and equity. As Klak and Rulli point out, the new localism is by no means a guarantee of equity in economic development. The efforts of the Caribbean states to attract export industries have degenerated into "hypercompetition" in which "success" is fleeting, the mobility of capital enhanced, and the

Table 10.5 Participation of Governmental and Nongovernmental Groups in Fiscal Policy Making

	1	*2*	*3*	*4*	*5*	*6*	*7*	*8*	*9*	*10*
Group	$\overline{X}$	*Finland*	*Japan*	*Australia*	*France*	*United States*	*Canada*	*Israel*	*Poland*	*Belgium*
Mayor[a]	61		56	60		60	66	93		
City council[a]	61	68	61	59	57	61	66	63		
CAO[a]	49			59		49	62	52		
Finance staff[a]	56	70	63	42		56	62	82		
Department heads[a]	58	74	58	51	57	58	61	64		
Public employee unions	49	70	49	33	44	47	44			51
Business	48	63	56	36	35	45	47		51	34
Elderly	46	62	51	38	52	40	37	27		
Neighborhood groups	43	57	49	35	45	40	35	25		48
Citizens	41	37	42	42	29	38	41	29	53	
Civic groups	40	71	48	16		35	25			
Low-income groups	36		39	32		36	38			
Homeowner groups	36	45	31	36	26	36	38			25
Minority groups	36	50		26		34	28			31
Taxpayer groups	29	36	40	18	15	30	34			
Local media	37	62	46	37		30	41	21	26	
Churches	38	30	28	25		23	26	30	34	43
Federal agencies[a]	35	56	58	32	28	24	26	51		
State agencies[a]	32		58	38		24				

a. Denotes governmental group.

benefits of investment minimal. In Hungary and Poland, the shift from a state apparatus dominated by the center to one with more regional and local autonomy has the potential to exacerbate regional inequities. Already, Péteri shows that patterns of privatization and the development of a market economy in Hungary follow distinct regional patterns. Although the major urban areas (especially Budapest) experience greater levels of investment and privatization, villages and rural areas show little entrepreneurial activity. This pattern is heavily influenced by historical patterns of development.

Equity within regions is also a concern. Jacobs and Hula argue that the privatistic development pursued by Birmingham, Baltimore, and scores of other Western cities has produced extreme inequities within cities. The Caribbean case is a further illustration of the trend. Yet this is not an inevitable consequence of the new localism. As argued earlier, there seem to be both "left" and "right" versions of horizontal restructuring. Municipal socialism, for example, was an explicit attempt to address distributional issues. In the United States, a progressive paradigm for local development has emerged that puts equity issues up front (see, e.g., Bruyn & Meehan, 1987; Clavel, 1986; Clavel & Wiewel, 1991; Goetz, 1990). The adoption of these different paradigms depends on a host of local political and economic conditions (Goetz, 1990), and can produce a wide array of strategies. For example, Chandler notes, rather ironically, that after the Thatcher government reforms of the mid-1980s, some local governments pursuing equity issues received greater cooperation from the private sector than from the central government or its locally created institutions.

Conclusion

It is my hope that this summary has highlighted the range of political and economic arrangements forming and re-forming as a result of the restructuring of vertical and horizontal relationships. The research reported here suggests great variation in the political responses of central and local governments to economic change. I argued in the beginning of this chapter that the question of economic activism in a global era reduces ultimately to a question of local autonomy. The weight of evidence provided by the contributors to this volume indicates a great deal of autonomy and the existence of important differences in local political environments regarding the relationship between economic and political restructuring. Indeed, Surazska argues that for Poland the question is not whether local governments can direct the local economy (a question of economic autonomy) but whether they will be allowed to by the center. The localized political arrangements made between center and periphery, and between the state and the market, are decisive.

Perhaps of greatest note are the tremendous experimentation and innovation occurring in local governance around the globe. While national leaders, at least in the United States, consider various versions of "reinventing government," it is clear that the redefinition of traditional rules of governance and political roles has been occurring widely for some time at the local level. With this process of redefinition, normative questions regarding participation and equity are also being reevaluated. The meaning of substantively new economic roles for local government is being worked out in a process of dynamic interplay among central governments, local governments, the private sector, and community-based interest groups. Changes in the scope of political participation and patterns of distributional equity are two important products of that process.

This volume illustrates the vitality of local politics and the importance of local variations even in an era of global economic change. The field of urban studies has begun to appreciate the commonalities of experience across national lines. Cross-national comparisons are growing, and the number of volumes containing case studies representing multinational settings is increasing. The chapters in this volume justify an extension of that comparative approach beyond Western cities to incorporate the commonalities across wide cultural boundaries, developmental stages, and political economies.

Notes

1. The Fiscal Austerity and Urban Innovation Project is an enormous data collection effort that currently has data on 12 countries. There are completed surveys of chief administrative officers in the United States, Canada, Finland, Norway, Australia, Japan, and France. There are completed surveys of mayors in those countries and Poland, Belgium, and Israel. Demographic data and fiscal data for cities have been collected in various degrees for those countries as well as for cities in Sweden and Denmark. The project was begun in 1982 and is an ongoing data collection and analysis effort.

2. As in the previous analysis, responses are scaled from 0 to 100, indicating an increase in importance as the scale value increases. In the category of shrinking the local public

sector are selling assets, cutting across all departments, cutting in least efficient departments, personnel layoffs, contracting out to the private sector, contracting out to other governments, shifting responsibilities to other governments, reducing administrative expenses, hiring freezes, reduction in workforce through attrition, reduction in services self-funded, reduction in services funded by central governments, elimination of programs, and decreased capital expenditures. In the category of increasing the size and scope of the public sector are pursuing additional intergovernmental resources, increasing taxes, increasing user fees, increasing short-term borrowing, increasing long-term borrowing, and regulating the rate of population growth. In the neutral category are lowering surpluses, deferring payments until next year, reducing employee compensation, wage freezes, reducing government supplies and equipment, reducing government travel, improving productivity through management techniques, improving productivity through labor-saving techniques, reducing overtime, joint purchasing agreements, and deferred maintenance of capital stock.

3. The mean responses are available for cities in the following countries:

	Government Reducing	*Government Expanding*	*Neutral*
United States	25.5	26.9	27.0
France	—	36.9	—
Japan	—	30.7	30.6
Australia	27.7	—	32.4
Canada	24.4	30.8	26.5
Total	25.9	29.7	28.6

4. The mean scores on the summary variable Horizontal broken down by country are as follows: Finland, 67.8; Japan, 45.9; Australia, 32.4; France, 22.7; United States, 21.2; Canada, 20.3; and Norway, 18.0.

References

Beauregard, R. A. (Ed.). (1989). *Economic restructuring and political response.* Newbury Park, CA: Sage.

Body-Gendrot, S. (1987). Plant closures in socialist France. In M. P. Smith & J. R. Feagin (Eds.), *The capitalist city.* Oxford: Basil Blackwell.

Bruyn, S. T., & Meehan, J. (Eds.). (1987). *Beyond the market and the state: New directions in community development.* Philadelphia: Temple University Press.

Clarke, S. E. (Ed.). (1989). *Urban innovation and autonomy: Political implications of policy change.* Newbury Park, CA: Sage.

Clarke, S. E., & Rich, M. J. (1985). Making money work: The new urban policy arena. In T. N. Clark (Ed.), *Research in urban policy* (Vol. 1, pp. 101-115). Greenwich, CT: JAI.

Clavel, P. (1986). *The progressive city: Planning and participation.* New Brunswick, NJ: Rutgers University Press.

Clavel, P., & Wiewel, W. (Eds.). (1991). *Harold Washington and the neighborhoods: Progressive city government in Chicago, 1983-1987.* New Brunswick, NJ: Rutgers University Press.

Giloth, R. P., & Mier, R. (1989). Spatial change and social justice: Alternative economic development in Chicago. In R. A. Beauregard (Ed.), *Economic restructuring and political response.* Newbury Park, CA: Sage.

Goetz, E. G. (1990). "Type II policy" and mandated benefits in economic development. *Urban Affairs Quarterly, 26,* 170-190.

Judd, D., & Parkinson, M. (Eds.). (1990a). *Leadership and urban regeneration: Cities in North America and Europe.* Newbury Park, CA: Sage.

Judd, D., & Parkinson, M. (1990b). Patterns of leadership. In D. Judd & M. Parkinson (Eds.), *Leadership and urban regeneration: Cities in North America and Europe.* Newbury Park, CA: Sage.

Le Gales, P. (1990). Economic regeneration in Rennes: Local social dynamics and state support. In D. Judd & M. Parkinson (Eds.), *Leadership and urban regeneration: Cities in North America and Europe.* Newbury Park, CA: Sage.

Leveillee, J., & Whelan, R. K. (1990). Montreal: The struggle to become a world city. In D. Judd & M. Parkinson (Eds.), *Leadership and urban regeneration: Cities in North America and Europe.* Newbury Park, CA: Sage.

Logan, J. R., & Swanstrom, T. (1990). Urban restructuring: A critical view. In J. R. Logan & T. Swanstrom (Eds.), *Beyond the city limits: Urban policy and economic restructuring in comparative perspective.* Philadelphia: Temple University Press.

Mayer, M. (1987). Restructuring and popular opposition in West German cities. In M. P. Smith & J. R. Feagin (Eds.), *The capitalist city.* Oxford: Basil Blackwell.

Markusen, A. R. (1989). Industrial restructuring and regional politics. In R. A. Beauregard (Ed.), *Economic restructuring and political response.* Newbury Park, CA: Sage.

Parkinson, M. (1991). *Strategic responses to economic change in European cities in the '80s.* Paper presented at the annual meeting of the American Political Science Association, Washington, DC.

Schneider, M., & Teske, P. (1992). Toward a theory of the political entrepreneur: Evidence from local government. *American Political Science Review, 86,* 737-747.

Smith, M. P., & Feagin, J. R. (Eds.). (1987). *The capitalist city.* Oxford: Basil Blackwell.

Index

About the Authors

James A. Chandler is a Research Fellow and Senior Lecturer in the Policy Research Center, Sheffield Business School, Sheffield Hallam University, England. He is the author of several books on local government, including *Local Government Today* (1991) and *Public Policy Making for Local Government* (1988). Most recently he edited *Local Government in Liberal Democracies* (1993). He has also published widely in academic journals, including *Political Studies* and *Public Administration*.

Susan E. Clarke is Associate Professor of Political Science and Director of the Center for Public Policy Research at the University of Colorado at Boulder. Her publications on local economic development policies and interest representation structures have appeared in American and European journals. She is currently completing research on changing interest representation mechanisms in eight American cities with the support of a grant from the National Science Foundation.

Edward G. Goetz is Associate Professor in the Housing Program at the University of Minnesota. He received his Ph.D. in political science from Northwestern University. His research interests are in the areas of local housing policy and politics, and urban economic development policy.

He has published articles on these topics in *Urban Affairs Quarterly, International Journal of Urban and Regional Research, Journal of the American Planning Association,* and *Economic Development Quarterly.* He has written a book on local housing policy in the United States titled *Shelter Burden: Local Politics and Progressive Housing Policy* (1993).

Richard C. Hula is Professor of Political Science and Urban Affairs at Michigan State University, where he also serves as Director of the Program in Public Policy and Administration and Associate Director of the Institute for Public Policy and Social Research. Prior to joining the faculty at Michigan State, he taught at the University of Maryland and The University of Texas–Dallas. He received his Ph.D. in political science from Northwestern University. His research has focused on broad issues of urban policy, and areas of interest have included local economic development strategies and the impact of private lending decisions on communities. His current research examines the debate over possible privatization of the U.S. public housing system.

Brian Jacobs received his master's degree in politics from the University of Leeds and a Ph.D. in public policy from Keele University. He has worked for both the Department of Architecture and Planning of Westminster City Council and the Overseas Projects Group at the Department of Trade and Industry in London. He is currently Senior Lecturer in Public Policy in the School of Social Sciences at Staffordshire University. He has written numerous articles and three books on urban issues; his most recent book is *Fractured Cities: Capitalism, Community and Empowerment in Britain and America* (1992). His main current research interests focus on ethnic minority issues and the international political economy of urban change, with a particular emphasis on comparative policy in the United States and the European Community. This links with his interest in the politics of urban restructuring and the changing nature of public decision making.

Thomas Klak is Assistant Professor of Geography at Miami University in Oxford, Ohio, and Adjunct Assistant Professor of Geography at The Ohio State University, Columbus. He has conducted fieldwork in several Latin American and Caribbean countries and has published papers on relationships among housing, urbanization, recession, and state policy. His primary empirical interests are in state programs and policies, housing conditions, the work experience, and regional variations in development within countries. His theoretical interests focus on alternative development policies for production and reproduction issues, the international division of labor, and theories of the state.

Peter M. Ngau received his Ph.D. from the University of California, Los Angeles, and is a Senior Lecturer in the Department of Urban and Regional Planning at the University of Nairobi. His main research interests are in development studies and planning, rural-urban relations, development of small and intermediate urban centers, and the role of human agency in development. He also teaches research methods and computer applications in planning. He has published works in *Development and Change* and *Economic Development and Cultural Change* and is coeditor of *Fieldwork and Data Analysis* (1981).

Dele Olowu teaches and conducts research in the Department of Public Administration at Obafemi Awolowo University in Nigeria. His research interests focus on local governments, public administration, and local development issues. He has published extensively on Nigerian local government and responses to economic crises.

Gábor Péteri is an economist and Project Leader at the Hungarian Institute of Public Administration (HIPA). He started his career at the Planning and Economic Department of the City of Budapest as a regional planner, working on urbanization problems of the metropolitan area. At HIPA he coordinated a two-year project on public education finances and participated in the preparation of the new act on local governments in Hungary. He is currently one of the coordinators of the East-Central European comparative project, Local Democracy and Innovation. He works on alternative service delivery methods and local economic development problems in the new local government system and is actively involved in municipal consulting. He is the author of *Finances of Public Education,* published in Hungarian, and editor of the Local Democracy and Innovation project's first publication on East-Central European local transition, *Changes and Events.*

Jamie Rulli is a graduate student in geography at The Ohio State University, Columbus, where he received a university fellowship to study development. His undergraduate degrees, from Ohio State, are in journalism and Latin American studies, and he is interested in writing about economic development issues for the general, educated reader. His regional interests are Latin America and the Caribbean and their relationships with the United States.

Wisla Surazska, M. Litt in political science from Oxford University and Ph.D. in economics from Wroclaw Technical University in Poland, currently teaches at the University of Bergen, Norway, in the Department of Comparative Politics. She has taught and conducted research at Wroclaw Technical University, Trondheim University in Norway, and Allegheny College in Pennsylvania. She has published on the theory of social choice and on politics and administration in Communist countries. Her most recent work, "Center-Periphery Studies in EastCentral Europe," deals with administrative reform in the post-Communist transition.